ESOTERIC PHYSIOLOGY

Also from Dennis Klocek and Lindisfarne Books

Sacred Agriculture
The Alchemy of Biodynamics

Climate
Soul of the Earth

The Seer's Handbook
A Guide to Higher Perception

Esoteric Physiology

Consciousness and Disease

Dennis Klocek

Lindisfarne Books | 2016

2016
LINDISFARNE BOOKS
An imprint of Anthroposophic Press/SteinerBooks
610 Main Street, Great Barrington, MA
www.steinerbooks.org

Copyright © 2016 by Dennis Klocek. All rights reserved. No part of this publication may be reproduced, stored in a retrieval system, or transmitted, in any form or by any means, electronic, mechanical, photocopying, recording, or otherwise, without the prior written permission of the publisher. This book is based on previously unpublished lectures given at Rudolf Steiner College Coros Institute in summer 2010.

Cover image by Leonardo da Vinci (1452–1519)
Le proporzioni del corpo umano secondo Vitruvio (c. 1490)
("The proportions of the human body according to Vitruvius")
Pen and ink and wash over metalpoint on paper,
Gallerie dell'Accademia, Venice
Design: Jens Jensen

LIBRARY OF CONGRESS CONTROL NUMBER: 2016948908

ISBN: 978-1-58420-192-2 (paperback)
ISBN: 978-1-58420-193-9 (eBook)

Contents

1. Inflammation, Sclerosis, and the Human "I" — 1
2. Catabolism, Anabolism, and the Life Forces — 20
3. Secretion, Excretion, and the Circulatory System — 41
4. Nutrition and Consciousness — 60
5. Glands and the Heart — 75
6. Nerve and Blood — 98
7. Esoteric Embryo — 120
8. The Senses and the Life Body — 140
9. Digestion and Emotional Life — 159
10. The Neurology of Imagination — 178
11. Emotions and the Will — 196
12. Remedies and Dysfunctions — 214

Bibliography and Suggested Reading — 229

Chapter 1

Inflammation, Sclerosis, and the Human "I"

One of Rudolf Steiner's foundational works on therapeutic work is *Extending Practical Medicine*.* In the beginning he posits a question, and I will start with that because it is probably the central question of the work I would like to do with you. If you get an infection, your system is designed to create an inflammation, to send organisms in your blood toward the invading pathogen to identify with it, copy it, spread the copies around to say: This is the bad guy; take care of things; mop up; and then start a healing process. In the beginning of inflammation, as the old people would say, *calor, rubor, tumor,* or *dolor. Calor* = heat; *rubor* = red; *tumor* = hardening; *dolor* = pain.

In the past, for a doctor who was studying, if you said *calor, rubor, tumor,* or *dolor,* the doctor knew you were talking about an inflammation. In the process of calor–rubor–tumor, inflammation leads to a tumor, forming a hardening. When the hardening happens and the pain comes, it's actually a sign that the whole inflammatory process is developing in a healing direction. Inflammation, even though it can kill you, is nevertheless a healing process. It's a natural healing process. *Sclerosis,* the opposite of inflammation, technically means something is being deposited or hardened, so that a tumor in the calor–rubor–tumor–dolor process is sclerosis, or a deposit.

In healing processes or in the general growth of the body, the formation of bone or the formation of a tissue from a fluid is a sclerotic process. We can call it the formation of tissue from fluid growth, but technically it is *sclerotic*. Sclerosis, which is considered to be a disease process, is therefore a natural process in the body. Inflammation and deadly infection, as

---

* See the bibliography for this and other books cited.

well as sclerosis, are all the body's natural healing processes. Steiner mentions this and asks this: Where does the illness come in? This is a really a very good question. The answer he gives is this: Illness is not in the body and its life forces. Illness is in the thoughts we bring to the body and the life forces. Illness is in the soul.

This is the picture I will to try to unfold in this book. We could say the body/mind connection, the soul, and the interaction of the soul forces—the forces of consciousness—interacting with the forces of life create a condition of tension because, in Steiner's view, the forces of consciousness, the force of the self, are the source of disease. This is a sobering thought, but it is at the heart of the revolution that Steiner brings to medical work. The pole of consciousness is the disease-producing vector—the disease vector, not the germ. The soul process creates a potential for the germ to find a milieu in which it can multiply. Thus, in germ theory, DNA and cloning, or whatever, all the evolution of thinking in terms of physiology—modern, contemporary physiology—doesn't touch the pole of consciousness except in an incidental way. For Steiner, the pole of consciousness *is the* disease-producing vector, the factor, and that is really different.

In my work, I started studying physiology when I first started writing my book *Seeking Spirit Vision* (1998), and I worked on it for fifteen years. Since then, I've continued to study physiology but never felt I could offer a course like this, because I didn't think that I understood physiology well enough. However, in the process of working with these ideas and working with biodynamics—cow guts and manure and gems, weather, color, and the other stuff that I work with—I began to see that there are certain patterns behind them all from Steiner's point of view, allowing one to enter physiology from a kind of mythic consciousness that has to do with the soul. This provided a way of understanding deep conundrums about such matters as diabetes and cancer. Why and how does diabetes work through families? What is familial inheritance? What does that mean? How does it work? Is it only genetic? What is a gene? I didn't feel that I could answer these questions very well without two keys. The first key is from *An Outline of Esoteric Science*: ancient Saturn, which is called the "fall of the spirits of darkness." * The second key is Steiner's idea of the "I"-organization (sometimes also called "ego organization"). In all of his medical literature he talks about the "I"-organization. I've asked doctors; I've asked therapists; I've asked all kinds of people: What is the "I"-organization? The answer is usually something quite vague, so I have had to live with this question.

As I understand it now, the "I"-organization has to do with the fall of the spirits of darkness. These two things got together in my head, and a light went on. These two pictures allowed me to understand this issue of consciousness as the disease vector: what and why it is and what it means. On the other hand, because it is consciousness it cannot be measured. Steiner himself explained in his therapy course that we should not try to measure the "I"-organization based on the results of some test we do, because we will not be able to do this, because the relationship between the "I"-organization and the physical body—a fundamental relationship—is magical. Although we cannot measure it, we can observe it. We can pay attention to it and see how it works; "by their fruits you shall know them." However, we will not be able to measure it, because it is a magical relationship, and the very essence of the magical relationship has to do with the fall of the spirits of darkness.

Thus there is a key, which I will present here, and I will keep returning to it so that, when I say "I"-organization or other spiritual-scientific terms, we will have a common language. If you understand how consciousness and the life forces, especially the elemental body—the elements in the physical body—interact and why the astral body and the life body are linked to each other, then biodynamics or color work or whatever will become clearer.

Steiner offers a motif in "the fall of the spirits of darkness in ancient Saturn" that starts with the "Thrones on ancient Saturn." This language

---

* This is one of Steiner's early, foundational works describing his view of cosmic evolution and the human constitution. "Ancient Saturn" refers to a stage of evolution in the cosmos and Earth, discussed at length in *An Outline of Esoteric Science* (chap. 4) and in a number of subsequent lecture courses.

comes from John Scotus Eriugena (c. 815–877), whose terms come from Dionysius the Areopagite in the first century. He didn't invent it but made it available to modern consciousness for understanding how the spiritual hierarchies relate to modern scientific concepts.

	Greek Hierarchies	Christian Hierarchies	Steiner's Hierarchies
1	*Seraphime*	Seraphim	Spirits of Love
2	*Cherubime*	Cherubim	Spirits of Harmony
3	*Thronos*	Thrones	Spirits of Will
4	*Kyriotetes*	Dominions	Spirits of Wisdom
5	*Dynamis*	Mights	Spirits of Motion
6	*Exusiai (Elohim)*	Powers	Spirits of Form
7	*Archai*	Principalities	Spirits of Personality
8	*Archangeloi*	Archangels	Spirits of Fire
9	*Angeloi*	Angels	Sons of Life (or Twilight)

The table shows an outline of the hierarchies in the old language of the Greeks and Christianity, and then the quality that they bring according to Rudolf Steiner, as he explains it in *An Outline of Esoteric Science*. The experience of those beings on ancient Saturn set the seed for the human physical body—the hierarchy of the Thrones is the hierarchy of will. As Steiner describes it, we could say that "will" is enthusiasm to be—warmth for being. He calls it the sense germ. It's a germ of what later would become a sense organ in the physical body. It is the germ of a sense organ because we receive all kinds of things in the world through our sense organs that tell us we are enjoying life.

I have a pint of Ben & Jerry's now, so I'm happy. I'm listening to Def Leppard, and I'm doing fine. Sensory experience brings *enthusiasm for being*. If we understand the spirits of will (the Thrones) we see that they bring great enthusiasm for being, but during ancient Saturn there was no real being yet, but just their existence as enthusiasm for being. There was a huge flux of warmth and will toward being, and that feeling went out from them. Then, from the being of that entity other beings arose who saw the enthusiasm and will and realized the wisdom in it. This will means something. Rudolf Steiner tells us they are the Kyriotetes, the spirits of idea, the spirits of wisdom. He says that they provided a plan for the enthusiasm of the will; they reflected it back to the Thrones.

A reflection—the motif of *reflection* is a real key, because it is the basis of our nervous system, which basically acts as a mirror. In physiology, reflection and mirroring are your failsafe, your integral pathways, buffers, antagonistic pairs of muscles, stimuli response patterns, glandular secretions, response secretions, and thalamic connections in the brain. All of this is mirroring. Right there in ancient Saturn, the mirroring process that the Kyriotetes reflect to the enthusiastic will of the creative beings to create a sense organ creates a plan, or the beginning of a patterning. He calls it "wisdom." Then other beings arose from that: the Dynamis, the dynamics. The dynamics showed up and said, between this great will and your reflection, there is a kind of space in between where something is happening. Let's call it movement. So the impulse is going out from the Thrones and is reflected back to them by the Kyriotetes. Now we have a plan, and not just a plan, but also a reciprocation like every other reciprocation in the whole physical body.

Every polarity, everything is reciprocal, which is why studying physiology makes you nuts. Just as soon as you think you understand something, the opposite is true, and then maybe we have to study this some more. This is because the wisdom of the great plan is to reciprocate, which creates the pattern of movement. Rudolf Steiner links the work of the Dynamis to what he calls the forces of life. Life forces move according to a plan, a wise plan. When those wise moving forces go out and in, and in and out, and out and in, we have the basis of what later will become an organism. Stuff goes out and stuff comes in. Mouth, anus, whatever you want to call it—that's the plan. In–out, here–there, center–periphery, periphery–center; blood to the capillaries, blood to the heart, blood to the capillaries, blood to the heart. The great plan is to reciprocate in motion, and when that happens it makes systems possible.

Then there is another hierarchy that recognizes and understands systems, and they give form to the systems of the life forces. This is the Exusiai. In addition to reciprocating forces, there are also patterns; there is the possibility of arising forms. Things have insides and outsides, then what is inside tries to get outside, and the stuff on the outside tries to get inside. We get sensory experience and inner responses and glandular responses to sensory experience, because this hierarchy that creates form, according to Steiner, also creates the potential for what he calls *psychic imagery*. Now we are no longer dealing with an original force but with a mirroring of a mirroring of a mirroring, and you know what happens in mirrors: pictures arise that are not the thing itself. This picturing consciousness that is arising is not the original thing—that is, the source of what will eventually become the inner soul response of consciousness that human beings have when we have a sensory experience.

What I've already characterized is "consciousness equals illness." This being that's created not only participates in this reciprocation, but it also has the ability within itself to have the experience of an image arising in it. That's the *Exusia,* the image of form of an original force, an original experience. The next hierarchy, then, is called in esoteric language the spirits of darkness, the Archai, and the spirits of personality. Here, not everyone, but a few on ancient Saturn received these reciprocating images. Steiner calls the images, which come from the cosmos, *imaginations,* a "code word" for the ether sphere, the forces from the ether sphere. He calls them imaginations.

The Archai are the first hierarchy to take the imaginations, and because the Exusiai create the potential for psychic experience with imagery, the Archai take that a step further, saying: Now that I have the inner experience of an image, I can see that there is something out there that is not me. I am reflecting the cosmos in my inner life; I have a small part of me that I take away from the cornucopia of forces, and I hold onto that. This holding onto the image causes a "Fall," or the potential for beings who have the senses to experience their separateness from the source of the imaginations. In other words, they have their own consciousness, and because they have their own consciousness, they have the doorway to illness.

When I discovered this, it was like rockets going off. That's it! When the Archai—not all, but some of them—have the experience of these imaginations coming from these exalted beings, they pull away from the wholeness, the *pleroma,* a little. In this gesture of "otherness," a gesture of the number two, we see the one who is not the original one but holds a little something off. We could say an "embryo" that is actually somewhat parasitical, or as Rudolf Steiner puts it, tumor-like. Overcoming a tumor-like force and bringing tremendous life—we could call this *gestation*. It is the great drama. The Fall in the Archai happens because they hold on to a portion of the imagination that allows them to see that the source of that imagination is separate from them. This represents a kind of wrinkle, because the part that separates from the creation is no longer under the recognizance of the original creator. It's slipped into a kind of funny little world, and the name of that world in Anthroposophy is the soul, or astral body. The seed of astral body is set by the Fall of the spirits of darkness, because they bring the potential for personality to develop.

The next step in this evolution takes place after the spirits of personality experience the images, or imaginations, outside of themselves. Now the Archangels have to deal with this. According to Steiner's *Outline of Esoteric Science,* the effort of the Archangels to deal with this inspires the Seraphim, the beings of love, the highest under the Trinity, to come down and join their forces, because the Seraphim and Cherubim actually have nothing to do with human incarnation. At the stage of ancient Saturn, they've already gone past it. The Seraphim and Cherubim are free and beyond it. Nevertheless, they are inspired by the action of the Archangels to deal with this will and to give love, which is their hierarchical force; they bring and give love to the Archangels, and from the love that the Archangels receive from the Seraphim they take this feeling of imagery of separating and have, as Steiner puts it, this inner experience: I exist because the image of the cosmos is coming to me. They love the image, and they turn it around. Their experience is this: God is in me, but it is different from having it just come in and go out like the other hierarchies; they can actually be aware of it. Their sense of how the imaginations come in creates consciousness in them that *God is in them.*

Next it goes down to the Angels, and they have to take this God-is-in-me experience, and their experience (they have to develop it somehow) of struggling with this inspires the Cherubim, the spirits of harmony, to come and support their work and to give them harmony. Thus the Angels feel the harmony of the inner picture. As Steiner puts it, they interact with the inner pictures and say: Not only is God in me, but I am also in God. I reciprocate.

Then there are us. This falling and redeeming is related to the issue of sensory experience and forming an inner picture. It's the key. When I have a sensory experience and form an inner picture, what has happened over time in the human hierarchy (us) is that, because of the falling the human being becomes more and more embedded in the physicality of the sensory experience, the physical nature of the sense world.* Instead of the sense world being populated by gods and goddesses (or *Kachinas,* or *Devas*), it's just stuff to modern consciousness, not elementals and Angels. Because our consciousness has fallen so far owing to the way our sense life and thinking operate, it's difficult for us to understand that hidden behind the sense world is the action of all the hierarchies and even the Christ-being in the Earth.

The next step in our evolution is to overcome the fallen nature of our sense life, to create in ourselves imaginations that we can give all that is fallen back up to the Angels, to the Archangels, to the Archai, and all way back up the chain. Rudolf Steiner said that the purpose of human evolution is consciously to give the creation back to the gods. The problem is that consciousness brings illness. As human beings, we try to bring the necessary consciousness to turn this around, which brings pain, anxiety, confusion, doubt, fear, hatred, and death—all the stuff we say shouldn't be part of the world, but is. We wonder who made this. We did!

---

* Here, we are talking about ancient Saturn, just spiritual. There is also ancient Sun and ancient Moon, where there are rebellions, and finally Earth evolution.

Adam and Eve—that is, all of us. We did it as a kind of schooling so that, as the tenth hierarchy, we could turn it around, with Christ as the teacher of how to turn the Earth back into a spirit. That's the great Rosicrucian teaching; Christ is now the spirit of the Earth, which means my sense activity has something to do with the work I need to do. This was the thing that made me realize that, in the fallen Archai, the retarded Archai, a seed was set in evolution so that the senses would be the catalyst, and that we had to develop thinking, or consciousness, to such a degree and experience the pain and anxiety of that so that we would be soft enough to turn around. This is Learning 101 for humankind. It's just the way we do it as the tenth hierarchy.

Here we see some polarities that may help in understanding physiology from the viewpoint of Rudolf Steiner:

Thrones: will-sensory germs
Wisdom: the wise arrangement of the physical
Motion: capacity for movement
Form: human body receives plastic form (etheric, or life, forces and psychic capacity(image forming or astral forces)
Archai: personality—similar to human earthly consciousness—develop sensory organs that treat images in such a way that they can perceive external objects; this inoculates humans with selfhood
Sons of Fire: Archangels receive light images from...
Seraphim (spirits of love): that give them picture consciousness
Sons of Twilight: Angels interact with the inner pictures, making it possible for the Cherubim (spirits of harmony) to inspire the Angels with dreamless-sleep consciousness

On ancient Saturn, with the fall of the spirits of darkness, there was a separation of imaginations as potential for imaginations as stuff, or phenomena. Imaginations* as potential, forces, will, and enthusiasm suddenly become "What is that out there?" Falling is the key to understanding physiology and how consciousness arises from glandular secretions, excretions, incarnations, and manifestations. The great polarity (we will keep

---

* On the esoteric meanings of *imagination, inspiration,* and *intuition,* see for example Rudolf Steiner, *A Psychology of Body, Soul, and Spirit: Anthroposophy, Psychosophy, and Pneumatosophy,* lecture of Sept. 15, 1911.

returning to this) is the first key motif: potential and manifest. All physiological processes occur as the tension between potential and the manifest.

Potential means it is limitless; it can become anything. *Manifest* means it is staying. Whether blood, lymph, digestive fluids, collagen, glycogen, or whatever, it is involved in those two polarities. They are what we call "absolutes." In alchemy, they correspond to *gravity* and *levity*. Gravity is manifestation; levity is potential; they are absolutes. In all physiology, they are absolute conditions. Gravity involves stuff that falls; with levity, stuff somehow rises. Of the two forces, science recognizes only gravity. However, all we have to do is go to the ocean and wonder why there are two tidal pools everyday. If we investigate this, we discover interesting facts about masses of gravity pulls, so that water on one side is "moving away" from the Earth, but this is because of—well, we have to figure it out. Levity is the realm of potential, life forces, activity, movement, creativity, will, warmth, enthusiasm, interest, potential. Work and light are levity forces; they have the potential to become manifest. Science tells us that everything around us is made of light. To be accurate, however, absolutely everything around us is made up of the corpse of light. That would be more accurate, but we tend not to believe that light is living; it's just light—cycles per second.

These are absolutes: levity–gravity, potential–manifestation—we could say substance, force, and substance. Force is potential, although in a sense it's actually manifest; it's actually there, having become something, so it is semi-potential forces. Steiner has the language for that. We have potential and manifestation as absolutes.

> **Cosmic etheric**: peripheral sculptural forces from starlight; these forces move from the periphery to the center in a planar movement.
> **Cosmic astral**: motions of the planets that interact with the peripheral forces and convert them into etheric formative forces; these have a musical quality and give rise to elements and the forces and laws of nature. These have been astralized to move from a center to the periphery.
> **Consolidating forces**: these take hold of etheric formative forces and the elemental patterns and convert them into substances, secretions, and excretions.

Here is the second "key motif": catabolism, inflammation, and anabolism. Catabolism is stuff breaking down into little pieces. In catabolism, the levity forces go away and substance falls out. Steiner calls it breaking down. This is what happens in consciousness; the potential forces are lifted, and stuff falls out. Steiner gives this as a picture of how the organs form. It is consciousness that is still up in the realm of imaginations, but then substance falls into the pattern and becomes a spleen or something. Then, once the organ is formed, the forces that fell out are attached to that substance and become the function of that organ. This is the principle behind biodynamic preparations. We can take a bladder and fill it with a bunch of flowers and later spray the substance on plants.

The key is to understand this. Suddenly the reason we use a particular organ and a particular flower starts to make sense. It is because the forces that form the organ are still there as potential, through the activity of the organ, even after I have removed it from the organism for a certain amount of time. The form of the organ can be linked imaginatively to plants that have similar organizational principles. When I bring the plant and the organ together, I get the yarrow preparation or the chamomile preparation. If you want a basis for this way of thinking in the work of Steiner, read Paracelsus. Steiner himself said that Basil Valentine and Paracelsus were the go-to people for him—helpers we could say. Many things that you find in Steiner's medical work can be found in the work of Paracelsus.

Catabolism usually takes place when a substance falls out of a fluid, because fluids, having the levity of life, carry substances. Fluids carry life. When a fluid carries life, Steiner calls that state *etheric*. The etheric body has fluids carrying substances that will eventually fall out of life into matter or even secretions. As the fluids move, substances move from substantiality into solution. In other words, the substances become more like activities and flow energetically along the energetic membranes of the fluid. That is the ether body. Those substances have to be rendered completely by the "I," otherwise we get reactions in the body through protein specificity. We reject the substances that enter but haven't been lifted to the level of life; our life body is geared to *our* life. When this happens—when a fluid comes into contact with consciousness in an organ—an organ simply is the center of a kind of consciousness that wants to get rid of it and let it pass by. Such biological actions are a state of consciousness of the particular organ. We call the different states of consciousness *temperaments* (we will look at these as we go along). The organ represents a kind of field of consciousness activity; when the fluid enters that organ, stuff falls out and becomes cytoplasm, and consciousness rises. That's a key picture from Steiner's work. This is an important key, because the astral body takes the life from the fluid. What falls out is substance, and the substance is in the form of the consciousness behind the organ, whether the liver, lung, or kidney.

The *how* is in the way the consciousness of my "I"-organization penetrates my physical body or doesn't. That's my main theme. I am trying to steer this discussion toward *anabolism*. This is when the levity forces unite with substance, dissolve it, and move it back into potential. Where it could become *my* thinking, it could become me instead of the tofu I ate. It's not as if your toes are made of tofu, however. The tofu has to disappear completely; the body will reject anything that resembles tofu. If you keep putting tofu in there, after a while it tries to reject some of it, and this would be called an allergy; your body is telling you, I want something else. Nevertheless, the body is patient; it sees that your soul doesn't seem to understand, so it sends a message in the form of pain, the great leveler.

Every reaction in the body has an element of catabolism, or breakdown, and anabolism, or building. Inflammation has a catabolic phase: calor, rubor, breakdown, destruction, phagocytes, leukocytes, and every other *-cyte*. Find it, kill it, eat it, break it down. This is catabolism. Then it's time for the breakdown to stop. If it doesn't, the breakdown starts to destroy your own organs. A doctor will diagnose this as an autoimmune response. When your own immune system starts to eat you, it is because your consciousness has not yet understood the fact that you keep repeating something that it is not happy with. After a while, the killer cells decide that they are just going to kill everything. That is autoimmunity. It is arthritis, Celiac disease, diabetes, lupus, and many others. It doesn't always have to involve substances taken in; it could be a certain state of consciousness that is repeated. We know that consciousness is the source of illness. Thus, whether it is inflammation or sclerosis, it involves a combination of these catabolic and anabolic forces. Anabolic processes tend toward building up—dissolving and building up. *Dissolving* means returning to potential so that other things can be built from it. This means dissolving something in its current form so that the ether body can build it into a new form. This is anabolism: buildup.

Catabolism occurs when something is broken down into little pieces. Each little piece becomes something that no longer retains the wholeness of the life body to support it. In that process, however, what is liberated is levity force, and then we get to the great issue of the "I"-organization. Therefore, in the fall of the spirits of darkness, the Fall took place in the hierarchy of the spirits of personality, the Archai. In the Fall the Archai received an impulse that triggered the Fall from the spirits of form, the Exusiai. The Exusiai formed an image of the inspired beings of motion. They passed this capacity for forming an inner picture to the Archai. For the Archai, the form of the future human physical body became an inner picture for them—so much so that they saw there was a separation between themselves and the form of the inherited picture of the potential physical body.

The Archai were the first to see that they were separate from the world. This was the Fall of creation away from the creator and into the direction of beings aware of having a self that is separate from creation. The Fall had to be redeemed so that, eventually, the Archangels, and later the Angels, had to receive the Fall but could interact with those pictures to change them. Instead of just perceiving that in my sensation the world is just out there, it can happen that the experience is transformed into my ability to perceive what is happening within me when I see something out in the world. To redeem the Fall, there needed to be a flow and reciprocation to it—the experience of something out there that enters when I am sensing. This can give rise to the experience that there is something out there that has a life force of imagination and potential that is doing things in my organism when I am sensing. Normally, the world is filled with forces, but I see stuff. A plant is growing, but I don't see the growth itself; I see the corpse that it has made. That's my dilemma as a human being. If I wish to change that, I have to see the *growing* rather than what has grown. To do that, I have to do something with an inner picture. I have to relate what is in the inner picture to what's out there, and relate what's out there to the inner picture.

Now the biggest picture—the best and most interesting picture that I know of these imaginations as a human being—is my body. My body is the best picture ever of the whole cosmos. Why? Because it's mine. What do you not understand about my body? It's a picture of the whole thing. It's a picture. The forces animating it are still in potential. I speak words, pulling them from somewhere, and I don't know where they're coming from, but I hope they are making sense. It's all in potential, and then it comes through the apparatus and becomes quack–quack–quack. Then you get it. That process is the "I" having to deal with the physical body, because one's "I"-being takes the forces of life and turns them into consciousness. The forces of life that you get from the cosmos—when you go to sleep and you wake up in the morning and your gas tank is full and they've cleaned the upholstery from last night's party—when you wake up, those forces are given to you, and what do you do? You start thinking about things. You may even become obsessive about them. It's as if we simply want to spend the forces. You've got them; spend them! If you don't spend them, maybe you're afraid there won't be more tomorrow.

That's what we do with consciousness; the thinking forces come in, and we change them into thoughts. Rudolf Steiner says that the magical thing that does this is our "I"-organization, because it was the picture that the higher hierarchies were sending toward the form of psychic imagery that came into form with the Exusiai. The goal was to contact a personality, and it's my "I"-organization that goes through these transformations. These imagination, or image-making, forces allow me to think, to form pictures in myself, and to manipulate them and do things with them—to build empires, to be either nice or ugly. These are polar options, because my "I"-being takes the forces of life and casts them down into the physical and then has to monitor how they're doing down there. The focus of the "I"-organization, according to Steiner, is a magical connection to the elemental world that arises between the physical world of the senses and the soul. We don't realize that, when we cast stuff down into our physical body, we are releasing forces of consciousness. We don't normally realize that, with our day-waking consciousness, we create a whole world of beings around us called elementals, which are really fragments of a whole personality. Elemental beings made by human beings are fragmented personality traits. It becomes our entourage of little beings with horns and hoofs and goat tails, and they're playing little pranks. The ancients saw these as characters that have the butt-end of a horse and the head of person—and those are just the nice ones. You don't even want to look at some of them, because they remind you too much of things that you'd rather forget.

Such formative activities are what we do in the "I"-organization when the creative forces of imagination fall into liver, lung, kidney, heart, bone, teeth, and soul. As substance falls, forces are liberated as states of consciousness that we can learn through meditation to work with (we will discuss this later). We aren't using our consciousness, so we invite these substances down to play. The "I"-organization encounters the physical body (see chart, page 16) and has to monitor the elements (fire, earth, air, and water), the fluid nature, and the substances falling in and out. It has to monitor all of this, and in the process it forms thoughts about the transformations. Science calls the realms where images become substances "glands"; esoteric science calls them chakras.

In the forces in our chakras there are a multitude of elemental beings. When the "I"-organization actually engages the physical body and really starts to go into the physical, the experience we have is what we call pain. When the "I"-organization penetrates the physical directly and stays there, that's pain. However, the purpose of the "I"-organization is to

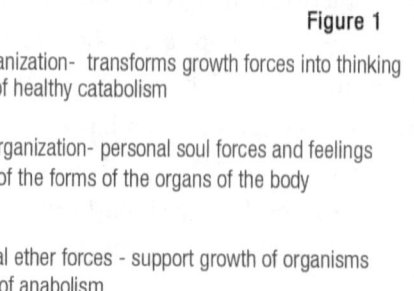

Figure 1

- "I"-organization- transforms growth forces into thinking source of healthy catabolism
- astral organization- personal soul forces and feelings source of the forms of the organs of the body
- personal ether forces - support growth of organisms source of anabolism
- physical forces- mineral formations-elements source of forces and substances

penetrate the physical. Who designed this system? According to Rudolf Steiner, the "I"-organization creates a will impulse, which stimulates an idea that leads to a motion. The "I"-organization grabs the physical body, it starts to move it in a particular way, and then the body is released. This is the way Steiner describes it. When it is healthy, the "I"-organization is continually grabbing and releasing, grabbing and releasing. When it is not healthy, it grabs and grabs and grabs, and this is what we experience as pain. The purpose of pain is to awake and alert the "I"-organization to what's going on in the physical, bringing whatever is going on in the physical into consciousness. As that happens, illness arises as a kind of gut check; this is how it goes in your soul.

These are the big bookends—sclerosis and inflammation. The "I"-organization works with the physical body to grab it and release it in health. If I need to deposit some substance for a bone, the "I"-organization grabs it as a deposit, and then out of that action consciousness arises. The bone material falls out; I get the collagen or whatever, and that is a deposition, but if I keep doing it, the bone becomes sclerotic. In life, however, when that process becomes sclerotic because I keep doing it, the result is an awakening in consciousness that we call pain. We say, I have a pain in my bone, which means that I now have a deep consciousness there. On the other hand, there is another form of consciousness between the astral body and the etheric body (see the previous chart).

Inflammation, Sclerosis, and the Human "I"

The astral body is where I have my soul experience, where I live in my soul. I am not always aware of it, but I'm sensitive to things— I flow in and out in my sense life in my astral body. This interacts with my life forces to create catabolism. This is where my astral meets my etheric in catabolism. My soul body interacts with my life body and I get the beginning of the deposition of forces; if it's balanced, things are deposited and then released in a catabolic gesture of dissolving. The catabolic dissolution is the prelude to anabolic building-up of new tissues. If catabolism and anabolism are all dancing back-and-forth, that's health. If my "I" organization is too strong it pushes my soul into my life force and my life force into my physical body. Then a response to this imbalance arises in my consciousness. That is pain. If my life body is too strong and pushes into my soul, I can become either numbed or hypersensitive.

To understand these dynamics you don't go to your *vade mecum*, but start by asking about the patient's lifestyle and about feelings and responses to things, because this is where illness enters the life processes. All of this is meant to give the big picture, and we will break it down into smaller pieces as we go along. However, we will keep returning to these themes, because I've found that when it gets complicated, we can stop and ask whether we are dealing with something anabolic or something catabolic. Is this going into potential or into substance? If the "I"-organization pushes too far into the astral or life body, we get anxious if it pushes in and stays there. If the "I"-organization goes into the soul or life body and then pulls back, unconsciously we become a little anxious, but then we bring consciousness into the process and feel okay. To do this, we have to learn how to move consciously. We can learn some form of art or learn how to tend a garden. Then the "I"-organization, the astral body, and the life body can all move down into the physical and then breathe out, and we are happy again. Then we have health.

Health and illness have the same root in the mobility of the subtle bodies to interact. We repeat acute inflammation that is healing until it becomes chronic inflammation and then a disease. This is why chocolate is both good for you and horrible for you. Both of those perspectives are correct. We think there's going to be some "correct" answer somewhere; we think that someone is going to discover finally the correct or universal answer to health. However, we are not going to find that answer, because both what is healthy and unhealthy are correct.

When I was in Delphi years ago, it was usually said that the saying, "Know Thyself" was carved above the shrine, but we had a docent who said that the sign actually said, "Know Thyself—Nothing in Excess." Know yourself, nothing in excess. The issue here is an excessiveness of health; we can be too healthy and become somewhat dull in consciousness. If we are too healthy, we don't become sick, so we cannot have compassion for those who are sick. We begin to think there's no problem. When we are too healthy, the life body creates conditions that lead to something that knocks us back a bit, so that we wake up. This is the interplay of the body consciousness and "I"-consciousness, or "I"-organization. These are the themes to which we will return again and again.

At this point, it would be useful to do a meditation. Imagine going inside the top of your thighbone. There is a place where all of the ligaments are attached, and then there is the head of the thighbone, which is inserted into the pelvis, and inside that bone is a major area where the bone marrow creates blood. There are other such areas in the pelvis and elsewhere, but in adults that area of the thighbone is a major source of blood cells. According to Rudolf Steiner, in ancient esoteric schools adepts "entered" their thighbones to understand the life forces. Imagine going inside the head of your femur. Enter it and just listen there and then inwardly do the eurythmy gesture for "A" (*Ah*).*

Going into the physical thighbone itself is catabolic—that is, it creates a kind of contraction in one's consciousness. Doing the eurythmy gesture "A" within the bone is anabolic; it is resonant with the formation of blood and life, which are expansive gestures. Thus, the gesture of contraction and forming substance and falling out is catabolism, the

---

* The "A" gesture in eurythmy involves lifting one's arms above the head and slightly forward, with the palms facing upward.

formation of dead stuff. Then we go into the bone as consciousness and we are in death; but we can bring it to life.

As Steiner puts it, we look at the physical bone as death; but to go into the bone consciously is awakening. In esoteric reality, the skeleton is an imagination of the "I"-organization. The picture behind that is—if you could see what the "I"-organization does—is a kind of musicality. If we had rib cage, we could hit it with a small hammer and it would sing. The sequence of ribs is the sequence of musical intervals of the formation of substance from levity force. This is the characteristic action of the "I"-organization that becomes embodied in that image of the musicality and harmonics of the physical body. When we go into the center of the thighbone, we go into one of the deepest mysteries of blood regeneration.

With each of the following chapters, we will add to this central imagination. We will go from the thighbone to the spleen to capillaries, and then we will move through the course of the blood in its journey through the body, doing eurythmy exercises to accompany the journey. We will build imaginations from eurythmy with the vowels and consonants and connect them to these great pulses in human physiology—catabolism and anabolism, the forces of manifestation and potential.

Chapter 2

Catabolism, Anabolism, and the Life Forces

We begin with some ideas from Rudolf Steiner to help us understand the concepts we are discussing. From *Extending Practical Medicine*: "Separation of higher level from a lower level is the first step in the formation of substances."* This is the process of separating an ether from an element. It's a very important idea, separating an ether from an element. Take, for instance, the presence of sugar in plant sap. The substances of carbon, hydrogen, and oxygen are kept in a state of levity by the action of the plant's life.

I talked previously about potential and manifestation. The life forces in a plant keep carbon, hydrogen, and oxygen in movement and activity and that represents a levity state in the fluid of the plant. It's the same for blood. So the fluid represents the levity force, where the substances are in potential. And when they're in potential, they are in the ether. When they go into manifestation they become elemental. This is a big key. In the ether realm, substances are potentials for particular forms, and in the elemental world substances are potentials of form that are starting to actually be manifest. That's the manifestation of the elements—earth, air, fire, and water. That's a very big idea.

The substances of the plant are embedded in the sap solution as the potential patterns by which the sugars manifest. In the plant, however, these substances are still engaged energetically in the life of the plant [etherically]; they are present as sentient or sensitive potential for attraction and repulsion.

---

* NOTE: Quotations of translated texts in this book have been revised.

Catabolism, Anabolism, and the Life Forces

These are ions—ionic exchange, gas exchange, pH, pH potential; hydrogen is just ionic potential of attraction and repulsion. I'm interpreting for you what says in the language of science, but does so esoterically. Energetically engaged in the life of the plant, they represent sentient or sensitive potentials for attraction or repulsion. Once the plant sap is taken from the living matrix of the plant, its constituents begin to move toward manifestation. The ethers that have animated the life and the chemistry separate from the sap, and the substances fall into manifestations, or images, of the energetic patterns.

This is the Fall, the creation of images. Sugar is an image of an activity, a sensitive activity, actually a mineral that is also connected to life. It's very unique in that it is a life substance that forms into a mineral, or crystal. The sugar falls into an image of the energetic patterns expressed in the living sap. Images are a primary way for potentials to become substance. Images are like a go-between. We can use the image and the form of the image to move substances back again into potential. That's the key to the Rosicrucian work. If I can find the right image, I can work with it meditatively and move it back into a potential state, because it has manifested from a potential state through an image. It's simple when you parse it this way. Steiner expects that you understand this. Whether you do or don't, that's his direction.

Now Rudolf Steiner's second concept: "Building form from separated substance is the second step." The first step coming from the fluid toward the formation of a substance is catabolism. The levity force of the form and the activity goes away and the substance falls. This is an image of the activity—catabolism, or breakdown. It is the action of the astral body; consciousness, levity, rises and substance falls, breakdown. Catabolism goes from potential to manifestation. In the second step, it has to return from manifestation into potential. The substance and levity forces have to be brought together again to build the new body. If everything in our body was catabolism we would probably not make it past the first week of embryonic development. Catabolism and anabolism are the tick-tock of the great clock of life. The difficult thing here to understand is that catabolism while being breakdown is actually the force that instigates building up of tissues. Dissolving supports building.

Quoting further: "Building form from separated substance is the second step. Once the carbon, oxygen, and hydrogen have separated from the ethers, they are like notes of a melody that is no longer being played." Beautiful—the notes of a melody that are no longer being played. You just have the notes without the movement between intervals. It's just the points, or substances, but not the activity; you no longer have the inner movement but just the substance. The valences, bonding angles, and molecular patterns of crystallization are fallen corpses of what once was a living, ordered, and unmanifested energetic entity. It couldn't be clearer, but when you first read it you don't get it. Now, with the background, you might begin to see that he's talking about the Fall, and now the resurrection, or the resurrection body, the phantom, the new human. We can think of the forces animating a good friend in contrast to the substances of carbon, oxygen, and hydrogen found in the tissues of their body. Which part of your friend elicits your friendship—the animating soul warmth and inner motion of the soul or the person's substances? Good question.

From Rudolf Steiner's lectures in *Education for Special Needs:*

> A phenomenon such as that of the astral body becoming congested in the lung occurs because the thought of the lung has not yet been properly integrated into the organism. This is the action of the "I"-organization to form thoughts from the life forces that support the formation of the organs. The "I"-organization takes life forces and forms inner pictures, the goal of which is to form organs in such a way that they are images of the imaginations that stand behind their activity. When that does not happen, illness results. (p. 65)

...when it doesn't happen *in the right way.* "All such phenomena are accordingly caused by the defects of thought. They are the result of our inability, as we descend into our organism, to gain control and be able to build it up a second time." Right to the point. This is anabolism, in which I have to take the levity forces that my "I"-organization formed into thoughts and somehow bring them into contact with substantiality again, to lift substantiality back to playing the melody and not just its corpse. I have to do that myself, because in the Fall I said, like a teenager: *Oh, thank you, but I think I can do that myself. The creative forces of the universe replied: Okay, go ahead and do that. We'll see how it works out.* We're still trying to work it out. Our work is to turn around what has fallen and bring it back into life. We do this in our anabolic side.

We take in protein from a source and destroy it catabolically, and then we rebuild in the image of the "I"-organization. This is digestion. We digest not only substances but also images, because images are the connection between the unmanifest and the manifest. Thus images often have a deep connection to whether we can digest or not. This is the source of allergic reactions, digestive problems, and so on. These thoughts will not allow us to incarnate the organs in the organism properly.

I have found that the only way I can understand these ideas is through inner pictures that move and flip-flop. If your inner pictures can flip-flop as they arise, you can understand difficult ideas from Rudolf Steiner. This is the nature of training to transform substance into life. It is the very thing required: creatively taking something that has fallen—transforming not only substances that have fallen but also our fixed beliefs and opinions. Fixed ideas have fallen out of the melody and fallen out of life. Then they represent our inability to manifest them in a proper way according to our destiny.

When we have a sensory experience we can bring a certain awareness to the qualities of that process. When Rudolf Steiner was building the first Goetheanum, people were carving curves to make the forms, and he said to them: Don't make these big florid curves. Make the curve as tight as you can. Otherwise, the curve will create astral fat. "Astral fat"—a great image. This occurs when you look at a curve in a sculpture. If it is a little too florid or sensual, it makes you feel a little queasy. Steiner always advocated making curves as tight as possible, meaning economizing on the curves so that they keep tension in them. You also don't want the curve to get flattened. Creating curved forms is a science as well as an art. The movement of your hands in rhythmic massage, for example, can be full of astral fat or not. A massage that's full of astral fat would involve superfluous, overly sensual movements that do not unite with the imaginations of the body. The pictures you hold while doing a massage are important.

Steiner told eurythmists in the therapeutic eurythmy course that, when you're working the vowels instead of consonants, it's very important to have something like a mental photograph of the form you're moving; don't just move. An old adage for eurythmists is to "leave your head in the costume closet, except when you're moving consonants." With the consonants, you need your head to form the picture as you move it, and the consonants will carry life, or ether, force. They are clear pictures he gives, based on this idea of the Fall.

From *The World of the Senses and the World of the Spirit*:

> Human beings were destined to be onlookers of themselves; it was not intended that they should live inside of themselves.... From Lucifer, human beings received the power of making the "I" preponderant over the astral body. This excessive egoism is the luciferic quality in humanity. If it had not been for Lucifer, human beings who would never have hit upon the ridiculous idea that they have an intelligence within them and that they themselves cherish thoughts within them. They would have known that the thoughts are outside of them and that they have to look on at their thinking. Human beings would always have considered and contemplated and then waited till the thought was given to them, waited until the purpose and meaning of the thought was revealed. Human beings would never have had the idea that they must connect all kinds of thoughts and form a judgment or opinion within themselves. This formation of judgments within ourselves, independent of revelation, is the luciferic nature within us.... The mystery of Golgotha was enacted to cure us of that singular arrogance and pride that manifests as a desire to comprehend everything by means of intellect. (lect. 3)

This points to the work we will do here in forming an exercise that follows the blood through the body. "Human beings were destined to be observers of themselves; it was not intended that we should live inside ourselves. Human beings received from Lucifer the power of making the 'I' have preponderance over the astral body." The human astral body receives images from the planetary realms. *Astral* is a code word for the movements of the planets. The "I"-being, however, was meant to see that movement and simply reflect it, but because of the Fall that didn't happen. Owing to the Fall, humans held over a little something from the experience of perceiving the astral realm, and perception started a long fall, and the great gods in the lodge of the gods said: Wait a minute; this is going to be an issue down the road. We need to turn this around. So they sent an envoy—the Christ—into an earthly body of flesh to teach the us how to deal with this, to recognize our need to turn it around. We need to look at this in a different way. I know you feel like you're living inside yourself, but really you are a limitless spirit. The only thing you don't understand is limitation; that's why you are embedded in this little flesh body. But Christ came from the farthest reaches to show us, in the body of flesh, that we can also see God. You can also be one with God in your flesh body, but it takes a lot of work. You have to love all of the other people to do it, not just tolerate them; tolerance is a good step, but loving them is the way.

Lucifer gave the human beings the power to take the "I"-experience that the Christ brought and say this is way better than learning all that hierarchical and planetary evolution stuff, because you can have the experience that you are above all the stuff that is given. The power to do that comes from Lucifer. It is the dominance of the "I" experience over the experience that I am living in astral world among astral beings who are creating images in me. Lucifer gives us that power when the "I"-being overcomes the astral body. "The preponderance of the 'I' over the astral and ether bodies is the source of illness." There you have it. My "I"-being takes the life forces and forms thoughts from them and lifts me through the thoughts above the astral body that rules the world—the harmony of the spheres and what regulates all of nature and the life of nature. My "I"-being says I have a deed to all of that real estate. I went down into a body of flesh, and I took a paper out of my pocket (now I take plastic out of my pocket) and moved its power from here to there, and now I own all that grass over there, all of the critters in the grass, and even what's under it. I can show you the deed. This is the preponderance of my "I" over the astral, over the sublime harmonies that make the grass grow. I don't say I'm taking care of the grass; I don't know who is making it grow, but I do say that that's my lawn. That's the preponderance of the "I" over the astral. And when that happens, the astral gets localized into *my* astral, and then we say *astrality*. But really, astrality is sublime when it's cosmic; it's the hierarchies singing of the planets' movements.

That's the cosmic astral. When we say astrality in a kind of disparaging anthroposophic way, we really mean my little chunk of astrality that Lucifer has allowed me to take away from the great astral with the preponderance of my "I" over the great astral; it's my little chunk of astrality I call "my soul." I can rob that little chunk of planetary forces from the big astrality, and that makes me a person, a personality. As I do this, I fail to realize that the process is a source of illness.

From *The World of the Senses and the World of the Spirit:*

> This excessive egoism is the luciferic quality in humanity. If it had not been for Lucifer, human beings who would never have hit upon the ridiculous idea that they have an intelligence within them and that they themselves cherish thoughts within them. They would have known that the thoughts are outside of them and that they have to look on at their thinking. (lect. 3)

Looking on at your thinking is meditation.

These images point to the work of somehow picturing my organism as a vehicle rather than seeing it as the best thing in the world and *mine*, and had better get out of the way of the Great Me. This is the illusion of a separate self. All of the mystery teachings have spoken of it since the beginning; they say that *the self is an illusion;* get rid of it. Because that illusion is already here, we have to do something about it. That's what you don't hear in the old schools, and the real issue in the new schools is, now that you have an abiding sense of yourself, you have to give that back to the gods to redeem what has fallen into the world, so that you

Catabolism, Anabolism, and the Life Forces

can have a sense of being separate from the world. No one can accomplish this task for us, because it is our job as the tenth hierarchy. There is no manual written on how to do it. We have the potential to do it because the Christ event made it possible for us to work on ourselves to the point where we understand that we need to do it. Giving up ourselves and throwing it away is not a good option today for Western people, because we are so infected with Lucifer. We have an infection called "I," but according to homeopathic laws, it is also the medicine. We're going to use the "I" in a certain way, and the work we are going to do now is with a meditation, picturing the blood's path through the body, starting in the marrow of our bones.

One exercise is to penetrate into the thighbone, because that is the source of our blood. We are talking about crazy numbers: 150,000,000 new blood cells every minute. Your blood is produced from the marrow of your bones—150 million blood cells per minute, a constant fountain of life. If that were all there is to this constant fountain of life, we would have complete anabolism; the ether body would rule, and we would be filled only with blood. Bicycle racers know this, so they store blood in a refrigerator, and before a race they inject more blood into their system so they have more potential for oxygen. And it's very difficult to track that with a chemical tracer because it's their own blood. So your own blood is the force in you of the ether body, and it is produced in the marrow of your bone.

When you put your consciousness into the marrow of your thighbone, you are placing it in a powerful center of your personal ether body, not the cosmic ether body, because the cosmic ether body comes from the stars. The stars come all the way into the marrow of your bones. Paracelsus says, "The human being is altogether a star." He called the human being an *Iliaster*, or star being. That star being lives by pulling forces from the ether world into the very death force of our bones. The most sclerotic part of our body is the bone, but it is permeated with life. It is a source of life, though it is the most sclerotic. This is completely the way thinking must happen in esoteric physiology. The deadest part is the source of the greatest life. And the greatest life is at the center of the deadest part. That's the rule. When I place my consciousness in my thigh, I'm asking the hierarchies: What is life eternal? We could call that the human Alpha. It's not just the thighbone that is the Alpha, but the thighbone is important.

When you are a child, almost every bone makes blood; you are like a blood-producing volcano as a child. This is the reason children can run around for hours without a problem. They are continuously producing life; practically all of their bones are producing blood. There is a time in embryology when most of the body is called a blood sponge. It's the cardinal system; it's like a sponge filling of the inside the embryo coursing with blood. Life. But in an adult, especially as you mature, parts of that blood-making start to get ossified, so the long part of your bones becomes crystallized and more mineralized. But the ends of the bones in an adult, where your growth was as a child, remains alive. So the thighbone is one of the largest blood sources. You have marrow in your backbone, you have marrow in your pelvis, you have marrow in your upper arm, at the terminus of them, but the largest amount of marrow is in the thighbone. Rudolf Steiner talks of penetrating the thighbone as a meditative device to understand the life body.

We will call that the "human Alpha"—the marrow where we have our ether source, or genesis so to speak. And the Omega, the "Z" or end, part is the spleen. The blood is formed in the bone, goes into general circulation, into the lymph circulation, and through all its processes; in the end, it goes into the spleen and is separated there and removed from general circulation. Human consciousness can then ask: Do I need this or not? This is used-up blood, so it must be removed from circulation. Damaged blood cells and all kinds of stuff are monitored in the spleen. Surgeons claim you don't really need it, so if it's not working it can just be removed. But when you don't have it, it's like having only four people on a basketball team. In a hockey game, it's the guy in the penalty box. You're not playing with a full deck without a spleen. It's the same with the gallbladder. The spleen and the gallbladder are connected to each other through this process of destroying the blood and forming bile. You don't need them, sort of, but it's really nice that you have them. The superfluous things, such as the appendix, have to do with this blood process of annihilating what was and getting it out of the body. If they don't do it, other organs will pick up the slack, but you're always a little behind.

In reality, the spleen is miraculous. Rudolf Steiner calls the spleen Saturn. What he means is that Saturn is listening because it is the farthest out and listening to the totality of how your astral body and "I" are using your life forces. That's what the spleen, Saturn, is listening to in the blood. Is this part of the grand destiny of your being, or has it run its course? It separates out the parts that can be refurbished and the parts that cannot be, and even the parts that can't be used to make something are used to destroy something else. That would be the bile made from spent blood. The bile goes from the spleen into your gallbladder and destroys the fat you eat, because your stomach cannot do it. Nothing is wasted, everything is economical. But there are various levels of consciousness surrounding these organs or involved in forming these organs. The consciousness of the spleen differentiates between "this is the end of this" and "that has further use." Then the blood that comes from the spleen is new blood, so to speak. It has been filtered, but it has to go back into the cycle and be revivified. In the figure, we have Alpha and Omega; we have the thighbone and the spleen (see figure 11, next page).

In the therapeutic eurythmy course, Rudolf Steiner brings certain pictures to the eurythmists that are very interesting. He talks about the consonants and says the consonants actually come from the etheric realm. We have the vowel sequence (L, A(h), O, U, M) and consonant sequence (T, S, R, M, A(h)). That's where I'm going here, because there are certain sequences of vowels and consonants that are used by physicians and therapeutic eurythmists when they want to heal sclerosis and inflammation.

Those are the two poles, sclerosis and inflammation. Sclerosis is the formation of deposits and a hardening process. This means sclerosis is the result of chronic catabolism that eventually creates deposits by

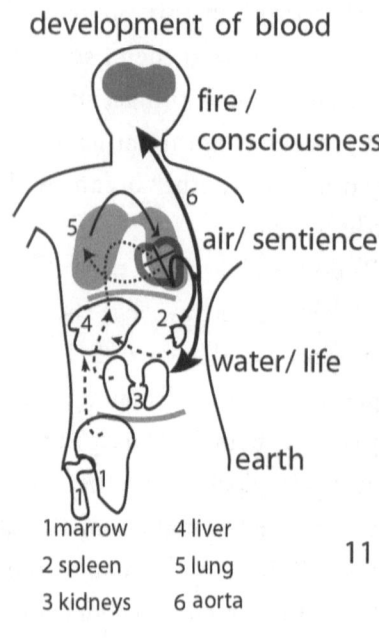

development of blood
fire / consciousness
air/ sentience
water/ life
earth

1 marrow
2 spleen
3 kidneys
4 liver
5 lung
6 aorta

11

breaking everything down into what an alchemist would call "dust." *Dust* is code for things that are so catabolized that they fall out of life and cannot be picked up by the life body. In catabolism that has not produced chronic sclerosis, substance dissolves to a point where it can be rebuilt into new life. Such catabolism is the basis for growth. Acute catabolism leads to the deposit of substance for new growth. If that pattern becomes chronic, the deposition leads first to a cessation of the dissolving forces, which in itself is not pathological. It just means that, if I want to form an organ from a fluid, I must first dissolve the original substance and then subject the dissolved substance to a sclerotic process to build the organ.

If my "I"-organization is healthy it catabolizes substance and then pulls back short of chronically fixing it. Benign sclerosis is the cessation of inflammation, and in it the substances get deposited but when the ego pulls back, the life forces come in and lift them through anabolism again into life so that the dissolved substances stay in contact with the cosmic imaginations. This is health; it is reciprocation. My "I"-organization, the thoughts I have, and the feelings I have cannot form the organ properly or reciprocally when my thoughts (and the processes that accompany them) become chronic rather than acute—that is, when they become habitual rather than circumstantial. They become opinions and judgments rather than interactive experiences. When they become opinions and judgments rather than interactive experiences, the quality of the thought and the karma of my own individuality in my "I"-organization create particular patterns in my thoughts that target one or another organ.

When I have habitual thoughts and feelings my "I"-organization does not pull back when it grabs into my physical body or into my elemental body. It does not allow the life body to reciprocate with a build-up because I'm only in a catabolic mode with my thoughts. My thoughts render everything into dust particles of information. I then have catabolic intellectual dusty, abstract thoughts. I have thoughts that are deposits of belief structures rather than creative original thoughts. But if I learn how to meditate in a way that I can allow the life forces to refurbish my thoughts, my dreams at night live in me in a different way in the morning.

This is the purpose of the arts. This is the purpose of the kind of food that the biodynamic method is supposed to create. It allows people to have imaginations. Steiner says this in his *Agriculture Course*. If the physical substances I take into my body are divorced from the life forces, it's very difficult for those forces to lift the substances because I have to overcome the chronic fixedness of those substance. The food does not contain the life, enzymes, and activity, and if my thoughts and feelings are habitual and I'm eating habitualized food, then I am likely to contract diabetes, atherosclerosis, or cancer. My depositing, repetitive, and fixed consciousness becomes dominant. In the distant past (Steiner talks about this), the predominant form of illness was inflammation. This pattern is represented by ulcerous growths and unchecked dissolution. Catabolism runs rampant because of a lack of reciprocal sclerotic healing. Membranes become infected because of rampant growth forces in the organism. These diseases are dissolving, excessively warm, and infectious.

However, as human beings began to think in a more personal and opinionated way, beyond the concept "the gods are thinking in me," diseases have moved from inflammatory, dissolving illnesses toward sclerotic, hardening ones. The diseases of choice today are sclerotic because of the overbalance of neurologically rigid, abstract ideation that has replaced hysterical magical beliefs. We can include viruses here, because a virus does not know if it's living or dead. The virus lives in its own universe, a little viral universe with its own rules.

Ninety percent of adults have Epstein-Barr in their system. It's not invasive but a normal garden-variety virus. However, when your thoughts reach a point where your life body cannot control them, then Epstein-Barr gets its hook in and replicates. These viruses have a stealth modality whereby the blood cells that work in your immune system cannot detect them. The virus can masquerade as a mineral. And the blood acts as though there is no problem. Even AIDS is this kind of depository disease, a sclerotic, fixed, mineral-like disease. The quality of modern consciousness forces us to think we have to make lots of do-or-die decisions and have to make them today and answer all the emails and so on—that consciousness drives us into a sclerotic frame of mind. We have to do this, have to do that, have to have an opinion about everything, otherwise we are considered bozos. When our minds contain nothing but repeated opinions, sclerosis overtakes the organ processes and, depending on the karma of your "I"-organization, some organ takes the hit based on your thought habits that prevent the organ from gaining access to life forces.

There are two exercises that therapeutic eurythmy offers. One is made up of consonants. Rudolf Steiner connects the forces of consonants to the zodiac, the ether sphere, or starry realms. Consonants provide human consciousness with a sense that life has meaning and that someone is watching the store. Although we have the key to the store, someone is watching the store, meaning there are forces beyond our cognition that bring life. Life and growth forces come from the consonantal realm. Consonant sounds create the sheath that contains life; they form the great womb of the human organism and hold potential; they bring the potential and move it in subtle ways—*b, t, r*. They form this outer vessel into which the forces of life can move; they create kind of a bowl, or sheath.

The consonant exercise is *T(uh), S, R* (rolling *r*), *M, A(h)*. If I make the eurythmy movements, if I make *T,* I move up and down at the same time. That's a plane. If I move *S*, I am moving side to side. If I do *R*, I am moving forward and back. So when I do *T*, *S*, and *R*, I touch all the planes in which a human being can live: up and down, side to side, and front to back. Put your consciousness into the thighbone and go *T*, *S*, *R*, *M*, *A(h)*. The *M* is the great balancing force; *A* is releasing as a vowel in this sequence of consonants. This is used to harmonize rampant astral situations—inflammations and infections and so on. It brings in the life force to calm matters. This is a big picture of who you are in the *T* dimension, in the *S* dimension, in the *R* dimension, and if you were mellow you would be like this, *M*, and then let it go: *A(h)*. Instead of actually moving it with our bodies, we'll form a little photograph of it and place it in the thighbone. We say to the hierarchies: I'd like a little of the cosmic etheric force in my center and source of my blood; this leads to anabolism, but if I have only anabolism I get rampant growth. Rampant growth leads to rampant depositing, which leads to something like a tumor. I don't want to do just one, unless I already have a problem, but if I have a tumor I would not want to do *S*, *M*, *A* twice. The tumor indicates too much growth, which needs to be loosened and moved. I need to get it to flow a little. I need to astralize it and get it to flow.

This is the purpose of *L*, *A(h)*, *O*, *U*, *M*, the vowels, which represent planetary movements, the cosmic astral. *L*, *A(h)*, *O*, *U*, *M* is used by physicians to work on a sclerotic condition, loosening and softening it. We go into the thighbone with *T*, *S*, *R*, *M*, *A(h)*—*L*, *A(h)*, *O*, *U*, *M*; thank you, thighbone. Then we go to the spleen with *T*, *S*, *R*, *M*, *A(h)*—*L*, *A(h)*, *O*, *U*, *M*; thank you, spleen. In certain traditions, going through the body in this way is called "the inner smile"; in some Eastern and Western traditions it's called "the body scan." We do a body scan, but instead of beginning mechanically from the toes and just moving up through, we try to make our scan follow the path of our blood.

I have been doing this exercise for a year and I can tell you it is really good. It allows you to refurbish yourself in the morning. Just start moving through it, through each organ. We do the Alpha and Omega to begin with and then begin moving through the capillaries, the liver, the lung, the kidney, the heart, and then up into the brain, nervous system, and so on. We follow the blood's path, from the venous structure to the arterial structure. As we do, we can bring in pictures of each organ from embryology to show what is happening. Eventually, you have a good inner smile, with which you inwardly connect with the feeling that your imagination is not merely random and that your "I"-organization can touch your organs without having to clamp down on them. Touch them, smile, catabolism, anabolism, and then move on.

I have struggled with this for about a year and a half, trying to determine for myself what this exercise really does. I've done variations of it in my course over the years, but I can say that this version of it is a good one and very safe, because you are just touching in and out. You have to just touch in and let go because of the "I"-organization.

You don't want to just go to your spleen and do *T, S, R, M, A(h)* forever, because then you are saying that you know how your spleen should work, which may already be a problem. This is just "a smile"; it does not make you a therapist or a therapeutic eurythmist. But the idea is that we want to get the "I"-organization used to checking in and then moving on. This, then, is the work in which we follow an astral–etheric pattern. The etheric is a source of anabolism; the astral is the source of catabolism in general. Catabolism is breakdown and dissolving for building up and anabolism is building up toward gradual formation, leading to breakdown.

In general, the venous circulation in the body (figure 11, pages 10 and 11) shows you the flow of the blood—there you are in the area of the pelvis and femur (number 1); down there, the marrow. And with that one it's going right up into the liver, but over there (number 2) is the spleen. For our work we're going to create a kind of gateway—I used to go from the marrow to the liver, but I found that the spleen felt lonely and left out at the end. It is a very important organ to bring into the smile. We will go from the marrow to the spleen and then through the capillaries to the liver. Following this, we come out of the marrow into the general venous circulation, which goes up into the heart; that's the anabolic side, the building side. Although the venous blood is filled with waste products, it's also where the forces of life enter, since the purpose of the liver is to take the dead stuff that the spleen lets go of and all the lymph and all of that flow from the periphery and make it live. In that venous circulation, especially in the capillary area, there is what is known as a secondary circulation.

There is a three-volume set on anthroposophic medicine by Husemann and Wolff, called *The Anthroposophical Approach to Medicine*, and in it there is a wonderful article about why we should not view the heart as a pump. It has to do with the way the venous blood is structured, the size of the particles in the venous blood, and their condition to receive the cosmic levity forces before going into the liver and allowing life and new protein synthesis to happen there.

Alpha and Omega—you go to and imagine the thigh. The thigh is the source of the blood, and the blood has cells in it that can become any other type of blood cell. This is very important, because when a blood cell morphs, it morphs into a blood cell that forms the basis of the immune system—white blood cells, killer T cells, all these specialized cells that learn to move through the blood and scavenge pathogens. Original blood cells have great potential to morph into any other, depending on where they find themselves in the circulation. In addition, there are certain organs such as the thymus and the spleen that act as "schools" for blood cells. In the thymus gland, blood cells become isolated from the regular blood, are introduced to pathogens, and have to react to the protein structure on the surface of the pathogen, most of which do not survive. Those that do survive remember that pathogen. Then those blood cells are released from the thymus into the general bloodstream with the ability, when a pathogen shows up, to go from being a cell with a definite cell wall and cytoplasm to becoming amoeboid meant to hunt and kill. They crawl through the blood vessels toward a pathogen because they sense that particular protein structure, having been schooled in the spleen. Those that survive leave; once burnt twice shy. This is the immune response.

Great magical things happen in the blood. Later we will look at inflammation in the blood in detail and at some of the patterns. This is a kind of consciousness under the charge of my "I"-organization, which observes how the life forces and astral body interact, how consciousness destroys life and how life and my ether body interact. My "I"-organization can do this because it dampens the life in the life forces to create a thought. This is an amazing picture. My immune system has this life that becomes sentient and knows its purpose. There are also organs of sentience—glands, the thymus—that are the center of the chakras, each having a particular consciousness. In those glands these little blood cells go to school and learn; they become licensed to kill. And when they do, they will just circulate around and when an invasion suddenly happens through a cut where something comes in or a virus comes in, those cells sense it in the blood, become amoeboid, and migrate to that site. It's almost like a dedifferentiation of a cell that has one purpose, to dedifferentiate, but then it becomes conscious. It is the "I"-organization that allows this to happen.

The L, A(h), O, U, M and the T, S, R, M, A(h) exercises are little microcosms of catabolism and anabolism. It is a little "breathing" process of breakdown and build-up. This happens a gazillion times each day in every interaction that every cell has with your blood. Oxygen goes from your blood and is deposited in the cell, and carbonic acid in the cell is deposited in the blood. The carbonic acid deposited in the blood is the basis of the future anabolic development, since it goes into the blood and relates to the phosphates and such there and starts building chains that will eventually be proteins. The oxygen leaves the blood coming from the arterial blood and enters the cell to create the ability of that cell to metabolize. These two forces of breakdown and build-up, catabolism and anabolism are involved in the tiniest little movement.

In the exercises of T, S, R, M, A(h) and L, A(h), O, U, M, we go into the thighbone. Concave is astral; convex is etheric. Springing from the curve is the etheric body, and hollowing out the inside is the catabolic side or astral. So in the form that you explore these two polarities.

Catabolism, Anabolism, and the Life Forces

If you cannot remember the vowel sequence, just remember something dissolving and then something growing. I have found that these sequences are very useful; sounding them inwardly is very useful. If you do L, A(h), O, U, M, there is a brief capsule of how these consonants and vowels work.

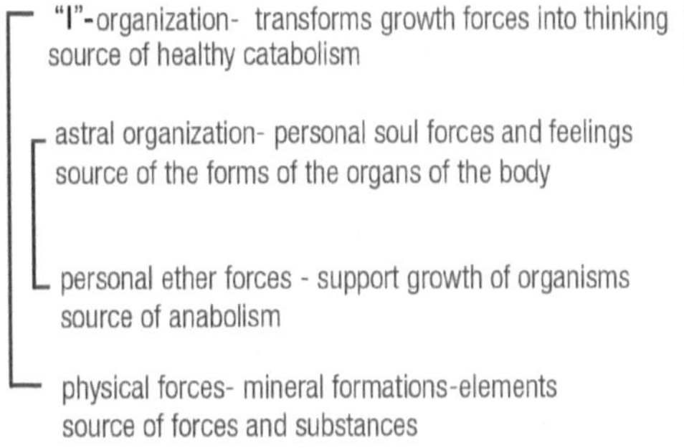

Figure 1

- "I"-organization- transforms growth forces into thinking source of healthy catabolism
- astral organization- personal soul forces and feelings source of the forms of the organs of the body
- personal ether forces - support growth of organisms source of anabolism
- physical forces- mineral formations-elements source of forces and substances

Figure 1 shows the interrelationships of the "I"-organization, physical forces, astral organization, and personal ether forces.

When a process starts it is calling into balance the exact opposite. There is no other way. Years ago when I was studying to write my book *Seeking Spirit Vision*, I read a lot on brain anatomy, as well as Rudolf Steiner's medical lectures, and I reached a place after a couple of years of reading and studying and trying to pin down Steiner on physiology. This is where I lost my bearings and got so far into paradox that I no longer knew which end was up. I got very angry at Steiner for speaking the truth. I thought he was doing a smoke-and-mirror trick. It was very tough for me, because I was also teaching here (at Rudolf Steiner College). People would ask me a question, and I would go to answer it, and two opposite Steiner quotes would come and self-immolate in my consciousness. I just couldn't answer, because the reciprocation of the two was causing free consciousness in me. But I have learned that those moments actually are growth moments. Studying physiology in this more creative way is a training in creative thinking; every polarity neutralizes every other polarity and makes it difficult to form an opinion that can't be changed. It's one thing to form an opinion and another to form an opinion that can't be changed, because as soon as you start forming fixed opinions you attract the need to change them. It is just human nature.

Thus, in physiology everything is an inversion. In alchemical language that's called the speculum. The speculum means a mirror. And it was understood in alchemy that there were two great forces in the world. One was the speculum, the mirror, where everything was mirrored in something else, meaning just the reverse. And if you couldn't achieve speculum consciousness, you were not worth your salt as an alchemist. That's why their language is so obscure. The obscurity is to winnow out the ones that don't have the requisite consciousness to receive the real picture, to tolerate the flip-flop of polarities. So the consciousness of holding polarities in the mind is called the speculum, the mirror; it means, what goes up comes down, what comes down goes up, what goes in goes out, what goes out goes in. What is above is below, what is below is above. So be it. And that consciousness is a new consciousness.

Our key motif is that all physiological processes occur as tension between two polar states of activity. One state is characterized as having potential to manifest; the other state is characterized as the manifest state. The potential state is composed of creative movements. The manifest state is composed of form. In physiology, form creates a condition in which what is formed initiates a process of breakdown. The breakdown goes through a series of steps in time (movements or processes). The movements lead to another buildup of potential.

The speculum is the first level of understanding, and the second level is called the monochord. The monochord is the arrangement of phenomena in harmonic sequences and intervals. Those harmonic sequences, the intervals of things, are the way mirroring progresses from one polarity to another. If I reach one stage of homeostasis, I'm guaranteed that the other side is going to call up the imbalance of the homeostasis, because homeostasis is not a chronic condition but an activity. Homeostasis means balance. But it is good balance only if it is an active balance. Balance is death if it is not active. The arrangements of polarities requires that somewhere the poles flip, and they need to flip; that is the speculum, or mirror. The particular way that the poles flip (the sequencing of the way they flip) is the monochord—this point is just a bit away from one pole and then the next point is just a bit farther than that one. The next point is even just a bit more distanced from the pole and so forth, until

Catabolism, Anabolism, and the Life Forces

I'm actually no longer moving away from the first pole but actually have begun approaching the other pole. This is systole and diastole in the heart rhythm. This is expansion and contraction.

What is the right phase for the poles of systole or diastole? Take your pick; either one is death. Heart expansion or compression? When the two are balanced with each other in alternating rhythms, you have the monochord, the EKG. There is alternating pulsing within certain parameters called ischemic response. There are parameters of the heart rhythms. The heart can be without blood for a fairly long time, but it has to be trained to do that. So in some cases, when heart surgeons are prepping a person who is going to be having open heart surgery, they insert a catheter with a balloon and shut off the blood supply for a little while and then release the balloon and then blow it up again. Each time they blow it up a little bit longer than the last to block the blood flow. The heart continues to beat without blood and then the balloon is deflated to let the blood flow through the heart again. Eventually they block the blood supply to the heart for a certain amount of time so that the heart is beating without blood. This is called *supply ischemia*. Through supply ischemia training, the heart is able to expand its rhythmic repertoire. They extend the rhythm, the monochord rhythm of the polarity of the blood flow, because they found that, if they trained the heart to tolerate supply ischemia, then if there is an event where the heart is without blood for a while in post-op, the heart can deal with the arrhythmia or fibrillation a lot better. Why? Because the heart has mastered supply ischemic response. A research surgeon in Ann Arbor told me that they were working with dogs. They could keep a dog without any blood in its heart for five minutes after building up the heart's tolerance. They could shut the blood off, and that dog would be fundamentally dead for five minutes, and then they let blood back in and the heart resumed normal function. The key idea is rhythmic training. This is remarkable. This is a technique used in surgery to help people who have heart issues after they have the operation.

The rhythm, the sequencing of the rhythm, the formation and letting-go are all part of our meditative work with the cycle. When you go into the thighbone you do the cycle—catabolic, anabolic, contraction, breakdown, buildup—and then let that image hover a while. The letting-go of the image allows that catabolic force to fill with life. We form the picture, open it up, dissolve it in there, and then we listen. That's how the system works, because catabolism and anabolism is how the whole system works. If we get into the practice of letting the image go and letting it hover a while, a new kind of consciousness comes into that space. So locate your thighbone; go into the very inside of it. Do the cycle and then listen. Imagine the blood flowing out of the bone, a great fountain of life. Now go to the Omega, the spleen. It's on the left side up under your ribs. Imagine the blood flowing toward the spleen. Do the cycle. Listen. Imagine the blood flowing out of spleen into the digestive area and then making its way out to the periphery of the body. Listen to the periphery of your body. And rest in silence.

Chapter 3

Secretion, Excretion, and the Circulatory System

We start by doing the exercise and add the capillaries, so we visualize the thigh, the spleen, and the capillaries, which are on or near the surface of the body. Imagine the blood coming from the spleen that has been cleaned, so to speak. However, it is not very alive, because the spleen has to do a great deal of scavenging and discarding to get rid of what you don't need. Splenic blood needs to be lifted into another modality. As the main secondary lymph organ, the blood coming from the spleen is moved into the lymphatic flow of blood and lymph that forms the basis for the portal circulation of the liver.

To start with the exercise, go into the thighbone. The shaft of the thighbone is mineral in adults, with only support tissue in it. The head of the thighbone contains trabecular (fine strut-like) fibers, with the marrow in the fibers. That's the living juice, so to speak, from which the blood comes. There is a kind of small chamber in there, where the elixir of life gets cooked, as in the alchemical wedding, to give life to the homunculus—the new being, the new king and queen. In that upper thighbone a web contains this miraculous substance, marrow. It turns out that marrow has a specific relationship to the substance of the thymus gland. This is interesting, because blood cells get trained to become part of the immune system in the thymus. The inner side of the bone and the thymus gland have this inner connection, so we could say that the thymus would be in the middle; Alpha is the marrow, the thymus is in the middle, and the spleen is Omega.

We will eventually get to the thymus as we go along, but right now we will go from the thighbone to the spleen to the capillaries, because as soon as the blood comes out of the spleen it needs to be enlivened. The spleen is on the left, up under the ribs. When that blood comes out of the spleen, it is clean but needs to be enlivened, or etherized. The liver is the organ that does this. It is on the right, but to get from the spleen on the left to the liver on the right, the blood has to go through lymph channels and then into the portal circulation of the liver. That is a major shift, because venous blood is on the way to building new protein in the liver, while the life forces of the arterial blood is being used up by the astral body, and carbon from cellular metabolism is being deposited into it as a kind of waste product, but it turns out that the wisdom of the Kyriotetes has made the carbonic waste products the basis for building new protein. The splenic blood cycles through the lymphatic system and the capillaries, where the blood cells have to fold up to go through the capillaries from the arterial side because the capillaries are so fine. In this process, the capillary wall is sensitive to the quality, difference, and potential of the arterial blood and the venous blood.

Muscular and nerve systems run back through the arteries to the heart, and then on the venous side there are no muscles since it's more of an open system without a pulse, with the blood being moved toward the heart through the movement of skeletal muscles. The liver, the destination of the venous circulation, is listening back through the blood to the spot in the capillary where the little venule is either open or closed. If it closes, no matter how much a blood cell folds it cannot get through. If it opens, the cells fold up and make it through. About halfway through the capillary, they queue up and cannot quite get through. Thus there is this play in the periphery of organs between the venous blood and the arterial blood, between catabolism and anabolism, or anabolism and catabolism. The venous blood is on the way to building up, and the arterial blood is being broken down. In the capillaries is an extremely delicate play of forces, and the regulation of that rhythm is known as a secondary, or peripheral, circulation.

The heart itself has nodes in it that sense the rhythm, pH, temperature, velocity, and viscosity of the blood. Those sense activities take place in the heart. On the periphery, however, the "I"-organization is listening to how the blood movements are playing out. The heart monitors the blood movements through the spleen, capillaries, and even the liver. The smallest parts of magnesium, potassium, and other mineral substances determine the fluctuations of capillary openings. A surging pulse created in the capillary net complements the heart's pulse. The shape of the heart stops the peripheral surge and is formed into rhythmic beats that are the true source of the pulse. The heart muscle doesn't just push the blood through the body.

According to researchers, there is a sensitivity between the heart receiving the peripheral surge and the shape of the heart chambers that form the basis for systole and diastole. This is an interesting idea. The heart receives the total accumulation of those movements and monitors it, but the actual pulse is instigated by a flow of blood from the periphery. Later, when we consider the embryo, we will see how the heart comes from the periphery and how the blood comes from the periphery in the yolk sac. This is an amazing picture. It points to the way the "I"-organization moves from the peripheral etheric into the manifest elemental world through what is called peripheral circulation.

Every exchange in which levity and gravity interact takes place under the auspices of the "I"-organization—every exchange. The "I"-organization, in Steiner's language, is consciousness that oscillates between unconscious and conscious as substances are moved, changed, deposited, released, broken down, and built up. The consciousness of the "I"-organization is there to monitor what, we could say, is the wisdom of the body. This is why Steiner calls it an organization, because the "I" uses the body for its sentience and consciousness and for its vehicle. The "I"-being lives in the warmth of the blood, but it uses that warmth to form an organization that allows the "I" to understand what is going on in the physical world.

Looking back to the ancient cosmic condition of old Saturn,[*] it all starts with the will, or warmth, sacrifice of the great creative

---

[*] See Rudolf Steiner, *An Outline of Esoteric Science*, chap. 4, "Cosmic Evolution and the Human Being," especially the section, "An Overview of Planetary Incarnations," pp. 125ff.

hierarchical beings traditionally called "Thrones," or spirits of willed enthusiasm for being. The "I"-organization operating in cells is a microcosm of this. The cell is enthusiastic for what's active in the blood—that is, the "I"-organization, in that enthusiasm uses the warmth of the blood to produce consciousness. Warmth can also be thought of as *enthusiasm for being*. However, we have to be precise when speaking here about warmth. Alchemically, we could say that the physical warmth we feel from a heat source is the corpse of living warmth, which is enthusiasm and love. We need dead warmth to warm our hands, but dead or manifest warmth always comes with the sacrifice of something that has to die and yield what was living as its corpse. The "I"-organization lives in the enthusiasm in the blood that expresses the will to be, to become. As our bodily systems exist, *becoming* means eventually to become a corpse, but to begin with you are not a corpse but a warmth organization, and then you are a light organization, and so on, and as you come down into the actual or manifest warmth, fire, and earth, you pick up the detritus of all the other corpses of the elemental beings and you build your body with them. But you forget that you are actually a star being composed of boundless enthusiasm, warmth, and the will *to be*.

In the capillaries we have this exchange between conscious and unconscious, between the astral body in the arterial circulation and the ether body in the venous circulation. At the surface of your body, there is the drama of the "I"-organization living in the blood and listening to how it is inside, but when we get out into the blood we get closer to the consciousness of watching ourselves rather than this is me here; get out of my way. (Rudolf Steiner talks about this luciferic aspect.) In the capillary circulation, think about capillaries and what actually happens there and imagine the blood moving along the surface of your body. Live into that activity; it is hugely alive. When you cut your finger where all the capillaries are, you'll see the blood, but if you cut yourself somewhere else, the blood is not as alive. A little cut on your finger will bleed a lot; cut yourself somewhere else and no

Secretion, Excretion, and the Circulatory System

problem. There is a consciousness in the capillary circulation that acts as a kind of polarity between that and what is inside your thighbone.

We will go from thighbone to spleen—Alpha and Omega, Joachim and Boaz, or however you want to look at it. As we approach, the first thing we encounter is the capillaries, which can be imagined as your starry realm. Your blood meets the stars at the little ends of your capillaries. The etheric sphere is incredibly active in the blood along the surface of your body; that's the imagination. Whenever I go into my capillaries, I picture a bunch of stars sending light into my blood that flows along the periphery of my body; it's a little bit like taking an etheric shower.

We are going from the center through the spleen and to the periphery. We keep adding as we continue, but at a certain point we can just start rolling. Go into the center at the top of your thighbone and through the cycle *T, S, R, M, A(h) – L, A(h), O, U, M*. Catabolism, and anabolism. Listen in silence for a moment. Move to your spleen on the left side under your rib cage and do the cycle—now listen. Imagine the blood flowing from your spleen out to the surface of your body, to the capillary net. Spread your consciousness along the surface of your body and listen to the blood. Go through the sequence *T, S, R, M, A(h) – L, A(h), O, U, M*. Listen in silence. Try to imagine the blood moving at the surface of your body just under the skin and then say the letters of the sequence and listen.

Eventually we will work outside the body, but we will go out through the head. We want to stay in the blood and eventually, when we get up into the head, the arterial circulation will take us up in there and we will dance around with the pituitary, pineal, limbic system, and so on, and then we go outside. When we go outside, there is another way that we will return down and go into the circulation of our hands. This comes later on. Go in and out and out and in, and then you have a complete cycle. This is where we are going. I will add organs and organ systems as we go along and we will investigate their embryology, so that we have a complete picture in the end.

Figure 5 shows a diagram and a circle with the word *Thrones* inside. This is an image of ancient Saturn. The enthusiasm of the Thrones, their warmth for being and will, radiates from them. If I am a being in the center and I radiate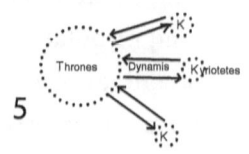
from me in all directions, you would see the form of a sphere. As Rudolf Steiner describes it, those forces radiate and, on the periphery of the Thrones, there were the Kyriotetes, the spirits of wisdom, that received that great gift of the Thrones' warmth substance and reflected it back to them. In a sense, it was a plan, or patterning, so to speak. Will flows out to the Kyriotetes, and then a speculum; it is reflected back from the periphery to the center. The movement out from the center out to the periphery and back from the periphery to the center is a patterning that later becomes organism. It is actually the exchange from a cell of what is in the center out to the periphery and then back again.

Everything in biology and physiology involves center to periphery and periphery to center. This is the basis for the motif on old Saturn. Then, in between we have the Dynamis, or spirits of motion, creating the forces of life that flow from center to periphery and from periphery to the center; this is another picture of the organism (this is in the upper part of figure 5). Finally, the Exusiai are added in between, so in between the Exusiai are the formative motifs of the flow—a picture of what will eventually be the form of a cell, organ, or organism. The other pictures in figure 5, starting on the left, show the development and ripening of an egg in the female ovary. The little circles of the cells are cells of the ovary, and the egg is a cell embedded in the other cells of the ovary; but the egg, when it is activated, has the potential to become thirty times the size of a normal body cell. There is a limit to the size of a body cell determined by *the law of minimal surfaces*. It takes a tremendous amount of energy for a cell to maintain integrity when it has such a large surface. There is too much going on, so the egg cell is the largest a cell can be in the body, approximately thirty times larger. This is interesting because, when the egg cell is fertilized by a sperm and division starts, the egg cell eventually divides thirty times without changing the nuclear material. This means that the potential of the egg becomes so large that it violates the fundamental law of the organism, and when it gets fertilized it begins to divide many times without changing the nuclear material in order to form the basis for a new organism. This is known as the *morula*.

You see in the diagram [page 1, figure 5] the ovary on the left with the egg in it, and the egg is thirty times larger than the cells around it. In the second stage are secretions from the woman's pituitary gland that flow around the egg in the ovary at the time of ovulation, causing an egg cell to begin a cycle of secretions, the method of communication in the biological spheres. Secretion is a kind of catabolic, digestive process. Rudolf Steiner pictures excretion and secretion as the way consciousness arises. This is catabolism in the astral realm. When there is a secretion, a dissolved substance is pulled from a fluid and concentrated so that it becomes a specific substance and there is, we could say as he describes it, a kind of freeing of the fluid, being relieved of the burden it carried by having that dissolved substance in it. Once the secretion of dissolved substances manifests as a particular compound, the deposition of the substance is accompanied by a lifting of consciousness.

There are all kinds of biological interactions that produce particular forms of consciousness through the release of secretions; they are neurotransmitters, hormones, and so on. Secretion of particular compounds results from a wide spectrum of stimuli, from putting something on your tongue that causes you to secrete saliva, to sexual relationships, to going to the bathroom. There is a catabolic process that is releasing something in a secretion or excretion. In excretion, the compounded substance is meant to be discarded; in secretion, the substance stays in the body to be metabolized so that its constituents can be reused. In this process, the substantial aspect of a secretion goes one way toward manifestation, and the fluid in which the substances are dissolved is released from its burden. Steiner pictures this process as releasing forces of levity that become available for consciousness.

In building a substance, there is a freeing of consciousness, but there is also a casting down of substances—secretion or excretion. This is why Steiner says that catabolism or the astral body or thought is the source of illness; when it becomes habitual, there is also a habitual kind of secretion. What if all you secreted every day were adrenaline? Perhaps if you work in an emergency room, this will get you through the day, but eventually you would burn out. *Secretion* is a kind of code word in Steiner's work. In *The World of the Senses and the World of the Spirit*, Steiner talks about secretion as the work of Lucifer. Originally, in sensory experience, the human being was designed by the hierarchies to have a sensory experience and to reflect that experience back to a divine source, but since Lucifer entered the sensory processes, we have a predominance of the "I" over the astral body. When the "I" acts on the astral body, rather than simply reflecting back, something is taken from the original body and held over to become concentrated, and it falls. It is an activity that becomes a substance. That is melatonin; it is progesterone, testosterone, adrenaline, a thyroid-stimulating hormone, and all these things. This whole realm involves activities that were imaginations fallen into actual substance, and the substance makes us do something.

In lectures given to Waldorf teachers, Steiner said that they really needed to understand secretion, because it is what you work with in children in the early grades and middle school. They are just total secretion beings; their consciousness is connected to their secretions. That relationship is called *the sentient body*. In Steiner's language, the part of the soul designated to monitor secretions is the sentient body. In that sentient body, when there is a secretion consciousness is filled with the activity and imaginations of the hierarchies behind the forces that Lucifer has caused to fall into a secretion. Where the hierarchies were giving those forces to us for life, originally we were just supposed to reflect them back. But because of Lucifer, we keep them to ourselves; we take a portion and hold on to it for ourselves. This holding on to the little piece creates a motif that becomes unnatural or un-godlike, a local god or Lucifer, instead of the Great Creator. This then leads to addiction and so on.

Secretion, Excretion, and the Circulatory System

All these patterns within us are linked to glandular secretions. Secretion is how the ether body communicates with the astral, or soul, body. When the "I"-organization moves the astral body, the feeling life, or the sense of "this is me here," arises in the sentient body. The sentient body is the part of the astral, or soul, body that monitors how the life body is reacting to sensory impressions. These impressions in the physical body affect the life forces in the life body (endocrine pathways); the life forces get squeezed out and fall into substances, like whey coming out of a piece of cheese, which is how Steiner sometimes describes it. The substance squeezed out is melatonin or adrenaline—that is, hormones and neurotransmitters. These substances result from the effect of a neurological experience in the sensory world stimulating the vascular side of a gland to secrete. The secretion then goes out through the blood in parts per million. This is when our heart is beating and we're running around looking for a mate, or whatever. It happens according to the imaginations within that hormone or neurotransmitter. Wherever we have the experience of secretion there is a kind of digestion in which a substance falls and a consciousness arises. In Steiner's language these patterns are all linked to secretion.

Returning to our egg, at a certain time of the month the egg is awakened by secretions from the pigment of the woman's retina, stimulating the pituitary gland, which stimulates an egg cell to start separating out its genetic material into polar bodies. This action creates secretions around the egg cell, causing its immediate environment to secrete fluids that pressurize the immediate area to force the egg to the surface of the ovary for eventual expulsion. Every biological event has to do with something touching something else and a secretion between them and then something new arising from that secretion. We could say growth and birth and all the other great mysteries tied to the human being are present in secretions, which create substance. With oogenesis, this process happens once a month. Under the influence of sensory perception of light, pituitary secretions stimulate a little cyst to form that in turn gets filled with fluid, which builds

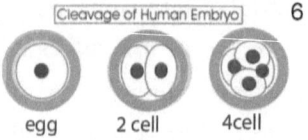

6

of pressure that pushes the egg to the surface of the ovary. The pressure of the secretion ejects the egg from the ovary for its meeting with destiny.

This is what you see in those diagrams. There you have the ovarian follicle; you have a secretion because of the interaction between the egg and the cells around it. The secretion forms a substance that creates a pressure that pushes the egg out of the ovary. The egg goes into the fallopian tube and forms a crown, or corona, around it. When the corona forms, the inside of the egg is chemically secreting to create a chamber from one end of the egg to the other, having to do with the nuclear material process of ejecting the polar bodies. In that chamber in the egg is the lone sweet spot where the polar body is ejected. That sweet spot on the egg is where a sperm zeros in. When the right sperm goes into the sweet spot, the way is already paved by secretions inside the egg so that the sperm genetic material will be absorbed, worked with, and developed.

Thus the egg becomes a zygote, with both female and male genetic material. It starts to divide repeatedly (see fig. 6). It is two days from the formation of the egg to the fertilization of the zygote. Within three days, there will be sixteen cells. The first two cells divide in the egg, according to the law of minimal surfaces, and form a plane in the egg that divides it vertically into two halves. People have followed the origin of that plane through to the formation of the actual embryo and find that the original direction of that first division is mapped to become the spine of the eventual organism.

From the first division of the egg, the organism is establishing itself—the head pole and the tail pole, and right and left. This takes place on day one of the zygote stage, when two daughter cells are formed. Then, twenty-four hours later, the zygote divides again. One daughter cell divides vertically, making two cells that define front and back, while the other daughter cell divides vertically along the same axis, forming left and right. Twelve hours later, the four daughter cells divide horizontally, creating above and below. Thus, by thirty-six hours the zygote of twelve cells establishes the orientation of the new organism in space.

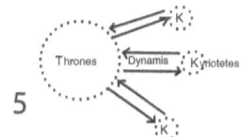

When the division reaches thirty, you just get close-pack. The zygote has just reached its limit, because it is thirty times larger than what it should be to begin with. Fertilization allows it to create within itself thirty more daughter cells that have exactly the same DNA as the first one, and there is no more mitotic or haploid division, just meiosis and diploid body division. It just divides into more cells with the same genetic material until it reaches thirty, and when it reaches thirty there is a condition called the morula.

You see the morula in figure 5. The morula forms what is known as the blastula from all of the mass of cells like close-pack, like a pomegranate. Some of the cells fall to one side as other cells start forming another periphery; they form another blastula with a blastocoel, and then you see the blastocyst and the embryonic disc. A process similar to what caused the egg to eject now starts again within the morula. The morula divides into where a hollow is formed, filled with fluid at one end and a cell mass at the other. The cell mass becomes the eventual embryo, and the hollow space filled with secretion becomes the sheath of the amnion and eventually the yolk sac and placenta. It just repeats a gesture similar to what it had in the ovary, but now we see the beginning of an organism.

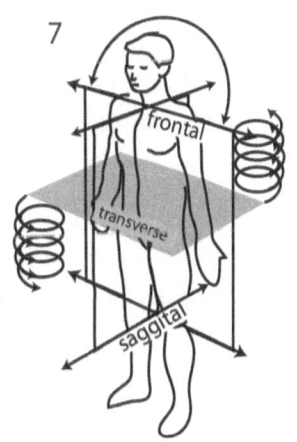

These patterns are a kind of recapitulation of the center-to-periphery-to-center that Steiner describes as ancient Saturn; it is the back-and-forth dialogue of old Saturn. Figure 7 shows the three planes, and figure 8 shows what happens when the blastula forms. We have one part that will become the embryo and another that will be the sheathes. In figure 5, you can see the embryonic disc and the blastocyst that embeds in the uterus. There's a hollow space where secretions form to become

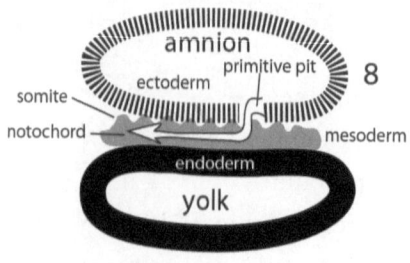

the amnion. That is the placental sheath; then you have a mass of cells at one end of the blastula that becomes the embryonic disc. We have the same type of pattern here—a hollow space with a mass of cells at one end, except now the mass of cells has actually been fertilized and is now a zygote. It is no longer just one kind of genetic material.

In figure 8, we see the yolk sac below and the amnion above. The amnion, the structure of the surrounding sheath, is composed of the ectoderm; *ecto* means out and *derm* means skin. This is the outer skin; in embryological development, the ectoderm gives rise to nerve tissue. Right from the formation of a blastula, as soon as it implants in the uterus, the disc divides itself into nerve tissue on one side and the yolk, or nutritive side, on the other. The yolk arises from the endoderm, which will eventually become the digestive organs. In human beings, there is a development in which the ectoderm of the amnion and the endoderm of the yolk touch each other and begin secreting into each other. The nerve side of the embryo and the metabolic side of the embryo touch and start secreting as cells move into the area, stimulated by the secretions and becoming what is called a mesoderm, or "middle skin."

The ectoderm becomes nerve tissue and much of the sensory tissue—sensory nerves, eye, eardrum, and so on; endoderm becomes the digestive organs; and mesoderm becomes everything else in between, including the blood, muscles, and the various life organs. The mesoderm, which leads to the life organs, arises through secretions shared between the nerve and metabolic poles. Very early in this development, there is a very unpredictable and mysterious event, the formation of the "primitive pit." The way embryologists describe it, in the ectodermal layer (fig. 8) a pit forms that penetrates down into the mesoderm arising from the secretions between these two layers. These are very intimate spaces, but

Secretion, Excretion, and the Circulatory System

the mesoderm is extremely active, eventually becoming all the blood vessels—not the blood, but the blood vessels, muscles, and life organs. All that carries life comes from the mesoderm.

The ectoderm, the nerve, descends into the mesoderm at the caudal end, or tail end, of the embryo. Embryologists describe it as a groove that forms on the surface as a hollow tube burrows through the mesoderm and cells dedifferentiate and migrate through the mesoderm toward the head pole of the disc. It's as if a finger were being pushed into a glove through the mesoderm, which descends down a layer as cells dedifferentiate and then migrate within the mesoderm layer to create a tube of cells that, once it reaches an organizing node in the center of the trilaminar disc, continues toward the head pole to form the notochord. In the diagram (fig. 8) this nervous tissue descends into the proliferating mesoderm and moves up toward the head, where it meets an area known as the organizing center. The organizing center eventually becomes the area of the pituitary gland.

The notochord is the final form of the nervous system in the invertebrates, which have a notochord instead of a spinal column. The notochord is on the front of their bodies, so when you touch a caterpillar, it folds around the front to protect its nervous system. They lack a backbone. The notochord is an inner tube, or groove. In Greek the word for groove is *psorat,* very close to *sorat,* a term Steiner uses to describe the Sun-demon, who engraves sensory issues into us when we incarnate. The notochord goes in and induces, through secretions from the ectoderm, the formation of the neural tube that becomes our spinal column. However, the first development is that the mesoderm creates a kind of hollowing-out, or astralizing, gesture in a recapitulation of the lower forms of nervous systems. Then the higher forms of the nervous system are induced by secretions from the notochord to form the actual backbone. In the embryo at term, the notochord becomes the content of the discs between the vertebrae. It goes away, but the place where the notochord occupies when it goes up to the organizing center, the space where it forms, eventually becomes the position of the autonomic nervous system.

The hollowing of a solid organ is the fundamental formative process of the embryo. Invagination, the hollowing of a solid organ, is the mark of astrality and the effect of the soul on the body. When a flat plane becomes concave, or hollow, this is the action of the astral body. When it is pushed out from the inside and made into a convex shape, this is the action of the life body. Convex and concave are the two formative gestures that form organs. The convex shape reveals a swelling from the inside, characteristic of the life body. The concave gesture is to hollow out and then seal off. This is a picture of the Fall of the spirits of darkness; something is taken from the original creation and sequestered in a being to seal it off.

Then it becomes possible to ask: What is that out there that my senses tell me *is not me?* Rudolf Steiner calls this astralization, the primary motif of organ formation. In an organism, a mass of cells is invaded and hollowed out, and the organ separates itself from the general circulation. Steiner calls this the action of the astral body. When it does, secretions form inside the organ and we have the catabolic force. A prime example is the development of the ectoderm moving in followed by the neural tube forming, which becomes the basis of the soul life of an organism. A general etheric mass is invaded and hollowed out, and we become *organi*zed. There are little compartments, and the organism rises in complexity, because, in a single cell, what comes in and then goes out is all; there is no mouth or anus, but only in or out. In a single cell, out is out, and in is in. They have vacuoles, or organelles, and such things to take care of. An actual organ seals itself off and then takes over a whole function.

Going through phyla and looking at them from simple phyla up to the more complex, we see that each phyla contributes a new prequel to an organ system that is incorporated into the next higher system. In the lower animals, there are all kinds of variations on digestive, reproductive, vascular, respiratory, and neurological systems until you get to higher organisms, whose diverse systems are stabilized and more integrated, and then vertebrates, and suddenly everyone has similar equipment. The

Secretion, Excretion, and the Circulatory System

organism becomes more sophisticated and, of course, more developed, but vertebrates don't have to compensate for the lack of a system of organs, which the lower animals lack the ability to organize.

You have the possibility of a much higher consciousness because you have all these different organs secreting. Once the body is built, so to speak, organ systems take care of things through secretions, which is how they speak to one another. As we move up through the phyla, we see increasing potential for consciousness, until you get to animals that have incredible consciousness and to humans who have self-consciousness. Self-awareness is the ultimate pulling away from the activities of the hierarchies, which is why the ancients warned against it: Don't go down there or you'll pay the price. But we did, so we have to pay the price. "Paying the price" means I have to penetrate through my separating consciousness back into the life organism and bring pictures into my life organism that don't destroy it when I think. Hence, the usefulness of esoteric physiology. I can actually move through the body, consciously bringing pictures and movements and trying to understand how the forces in the body are organized.

Our chakras are the new organs of perception. When Steiner describes chakras, he states the rule that half of their petals have already formed, and that the other half must be formed in harmony with those already formed. This means that there are certain patterns in the way the chakras operate, even in the patterns of consciousness associated with them that have to be incorporated into the new form. We cannot just completely do away with them because they are the hierarchies' gift to us; but the chakras, as they exist, do not include the problem of freedom.

When I am completely within myself, I have autonomy. The psychology built around autonomy means that a lack of autonomy in one's life leads to very deep dissatisfaction and eventually to illness. To many of us, autonomy is more important than relating to others. There are three psychological human needs: autonomy, competence, and relatedness. If any one of those three are missing, we are on a path of dysfunction. Autonomy is the most important one for people today. I can live in a system in which I have autonomy but competence and relatedness are not as present—a little like solitary confinement, which is a form of torture but as a human being I can exist if I have autonomy. If I have relatedness and no autonomy, I have a problem. I rebel against the collective. If I am very competent and I have no autonomy, we call that the "old knight syndrome." I'm working as hard as I can and doing a good job, but nobody appreciates me; I don't get recognition of the unique contribution of my individuality. Autonomy is linked to my sense of freedom as an individual. But autonomy does not mean anarchy.

Autonomy means that I can actually control decisions I need to make. If I lack that, I am on my way to degeneration. This problem becomes the role of the arts. If I live in a chronic situation of lacking autonomy, I have to find some form of autonomy in my life. I have to create images that allow me to build organs of cognition in harmony with the old, but including the whole issue of autonomy and freedom. Relatedness is the way the old world worked; you're related to me, and I'm related to you. I'm a butcher, you're my son, and so you are going to be a butcher. If you have a problem with that, move to another village. That's the old way, in which relatedness was the most important thing. Competence in slopping the pigs was not a big deal, and in the ancient world expressing autonomy was cause for exile from the collective. Now it's reversed. If you don't express your individuality you are marginalized. To a contemporary person, autonomy is the big issue, seen as freedom to express individuality. This means that, when I form my chakras for the future, I have to include the autonomy issue emotionally for myself.

The search for individual autonomy is the basis for most people's life decisions today. For my sons working world, especially computer work, if you don't change jobs every two years people wonder what's wrong with you. You are seen as unwilling to rise to the top. But when I grew up, if you changed jobs every two years people wondered what was wrong with you. You can't keep a job? A thirty-year pin was seen as an accomplishment. Today a thirty-year pin is the kiss of death. This is a symptom of the change in how the new chakras are asking to be built. The hollowing of the organism has created the potential obsession with autonomy. I have to be careful that my autonomy does not negate relatedness, however, because I also need relatedness. I also need to be on guard that autonomy does not negate competence or suffering incompetent fools. This is important, because incompetence is on the rise; it is epidemic, because the unconscious drive for autonomy is not balanced in the soul life of many people.

As an organ, the spleen contains white blood cells, but most is composed of red blood, so it is actually a reservoir for red blood cells. However, when the red blood enters the arteries in the spleen, the white blood cells that completely cover the arteries have been trained by the thymus basically to taste the blood; that is their function. In the spleen, they are isolated from the actual blood in that process—hermetically sealed off because they need to be isolated, otherwise they will react all the time and secrete into it, and the spleen would be unable to function, saying this one works, this one doesn't. Eventually when the blood exits the spleen, only a very few of these white blood cells are left. There is a kind of filtering process that is the function of many organs. The liver does it; the kidney does it; even the lungs do it. The life organs, because they are separated out, each do it differently, but they all have to do with some system of fluids moving in, with needed parts being separated from those not needed.

Each of the life organs, through its catabolic-astral nature, is a source of a kind of consciousness. The discriminating consciousness is called melancholia in the lung. The discriminating liver gives rise to phlegmatic consciousness. The discriminating kidney gives rise to hyper-vigilance, or sanguinity. (We will work with this later as soul motifs of the life organs.) Because it does this filtering, discriminating process, each organ casts something down, either a glandular secretion or the excretion of an organ such as the spleen. The spleen excretes blood that has been processed, and a kind of consciousness comes from that. The "I"-organization monitors that consciousness, as is its function. Over time, the "I"-organization is in touch with all these different levels of consciousness, and this becomes "me." That's my inner life, or work.

Goethe said that the human being is all of the animals combined, guided by an Angel. A cow has a very phlegmatic temperament, because it spleen is huge, its liver is gargantuan, and its digestive system is enormous, so its consciousness is that metabolic–spleen, liver kind of consciousness. It is a digestive consciousness. If that were us, we'd have the problem of chewing breakfast as a career. Cows chew breakfast all day because the cow over-soul is not *in* that cow. The over-soul carries the species' consciousness. The cow over-soul lives in the realm of archetypes. Individual cows, while living in a physical body, do not have direct access to the cow over-soul. The individual cow gains direct access to the cow archetype consciousness by dying. However, we have the potential to access our "I," or over-soul, while living in our physical bodies. This creates a tremendous death process in the human body that the cow does not have. This death process is called self-consciousness.

In a healthy human being, "I"-being, with its "I"-organization, holds all of those lopsided organs in balance. It monitors whether the liver is doing this and the kidneys are doing that, and it integrates the organs. That's the purpose of the "I"-organization, which takes life forces from the organs to integrate them into a centralized "I"-consciousness. This is done so that I don't go chronically into one organ to the exclusion of the others. This, then, is being temperamental, because when I am temperamental I am prevented from accessing my "I"-being. If I spend all of my time in my liver, I'm not truly human. If I spend all of my time in my lungs, I'm not truly human. I am a temperamental human, but not a full human being. This is our dilemma if we do not master our temperament.

As we develop in time, the liver is rounded off at seven, the lung at fourteen, and the kidneys at twenty-one. These are the developmental stages at which I try on a temperamental disposition because that organ is dominating the developing consciousness of the "I," or true Self. In his educational work, Steiner gives this developmental sequence as a picture.* Educators need to know that in a kindergarten we are really dealing with liver consciousness, in the third grade with a lung consciousness, and in high school with kidney consciousness. These are different parts of the astral body that, by the time a person turns twenty-one, should be mastered and integrated. Because of karma, my "I"-being comes predisposed to a particular family that might all trash their livers every night at dinner. Then I grow up with that disposition. Or a family in which everyone is hyper—kidney-type people who are always in a hysterical mind-set. They get into their kidney consciousness, and they see things as hysterical, and when the ego comes on board, they can't overcome it. Karma is based on the organ that dominates through temperamental disposition rather than the thought processes of one's "I"-being.

---

* See, for example, *The Child's Changing Consciousness: As the Basis of Pedagogical Practice*.

Chapter 4

Nutrition and Consciousness

I used a few books by Rudolf Steiner to prepare for this topic. *An Occult Physiology* (sixteen lectures) is amazing but very dense. It is a good book to read if you have some background in the language of physiology. *Man as a Being of Sense and Perception*, a booklet really (three lectures), is also very good. *The World of the Senses and the World of the Spirit* is a "huge" book (actually it's small, with only six lectures) and great study material for colleagues in Waldorf schools. Rudolf Steiner's *Education for Special Needs* (twelve lectures) is the basis of the Camphill movement and gives very clear pictures about pathological conditions, especially ways that emotional conditions relate to physiology. It is very good material for those who work with special needs children or adults.

The flagship of anthroposophic therapies is *Extending Practical Medicine: Fundamental Principles Based on the Science of the Spirit,* on which Steiner collaborated with Dr. Ita Wegman. It is an extremely dense book; even doctor and physical therapists find it challenging. It is difficult because every paragraph is like a whole chapter of any other book. You have to read and reread it, but it offers the fundamentals.

There is also a three-volume work by Friedrich Husemann and Otto Wolff, *The Anthroposophical Approach to Medicine*, starting with the first and second stages of life, developmental issues and diagnostics skills for physicians, developmental disorders in childhood, and a big chapter on inflammation and sclerosis as basic tendencies. If you want to understand that polarity, read that. Hysteria and neurasthenia as the two poles of inflammation and sclerosis, and then fundamentals of a biochemistry and physiology of disease processes. How biochemistry and pathological physiology relate—this is plant remedies, protein metabolism, light bearers in us, disorders of light metabolism in human beings, and stuff like this. It is an amazing set of pharmacodynamics, human and plant, the rhythmic system, typical healing plants, a capillary dynamic blood test, blood crystallization as a diagnostic technique, and so on—and that's just volume 1. The third volume is like a *vade mecum*. It tells you if the patient has this, so try this. If that doesn't work, do that.

Then there is Wilhelm Pelikan's two-volume *Healing Plants: Insights through Spiritual Science*. It is very useful for research into plant physiology and the production of healing substances.

Next is *Functional Morphology*, the great deed by Johannes Rohen. He was a professor of physiology at a teaching university in Germany. He is a deep anthroposophist, but in that milieu he was in the closet but kept writing his wonderful take on physiology and anatomy and putting it in the context of Steiner's threefold organism. This is his masterwork. It is down to mitochondria, in the detail, but always in a threefold context so you really get the picture. I couldn't have done this course without Rohen, because it allowed me to answer my questions and correct my misunderstandings, especially for the biodynamic work. Rohen puts Anthroposophy into the context of morphological development to see the themes in a given organ system; you see how they unfold and where the pathology comes in. He doesn't give diagnostics, and there is no pharmacology or anything like that in the book, but if you are used to reading *Grey's Anatomy* and trying to make functional physiological sense of standard anatomy, this book is a lifesaver.

Reading *Grey's Anatomy* is like reading a manual on how to take apart an airplane engine. But *Functional Morphology* puts anatomy into the context of the dynamic of organ and tissue morphology, giving very clear pictures. It is not overdone and it is not underdone, it's not gussied up. He doesn't avoid the big things such as "the heart is not a pump," but he gives two sides: this is how it is from here and this is how it is from there. He does it all as a brilliant morphologist and physiologist.

You can read some of these books and then return to Steiner and better understand what he is talking about. In this course, I am trying to give very fundamental pictures from an esoteric point of view that can dovetail with facts from physiology. Steiner had a good grasp of physiology, but most of his lectures on that subject were for doctors, and he assumed they knew physiology, so he gave them the full course on the esoteric work, and linking the two is not so easy, especially if you do not know the physiology of his day. MRIs and other technologies have given new pictures to blood flow, neurology, and the embryology. Steiner's references are to what was known at the time, but his insights are not limited to that time; they come from esoteric consciousness and have value far beyond the data of a particular time.

It has to do with the action of the "I" and astral body and the esoteric premise for human development. It can be very confusing, and you get almost angry because you might read something that totally contradicts what Steiner presents. But the contradictions are only in your mind, because your consciousness is not large enough to accept the esoteric truth. I am speaking from many years of experience and not having adequate consciousness to comprehend the truth. The reason I spend so much time studying physiology has to do with something Steiner said—that if more people studied and understood physiology, the social aspects in Anthroposophy would be much better, because you would not be dealing with a creation of the gods. When you study physiology like this (I'll go out on the limb here), there is no way one could be an atheist, because it is just too mind-boggling how incredibly and beautifully the living systems of the human being function. There are just too many miracles to deny divine intelligence.

The study of human physiology allows me to ask a person: How is your liver? Instead of, What's the matter? I can ask what time do you eat at night, rather than going for the jugular with my opinion or my answer. And so it has been a survival skill for me for years to try to study physiology as a teacher of adults. Because they all come with a particular twang in their physiology, and that creates a soul mood that gets put on my roster. And then I have to treat that as a child of God. It's easy when they are five years old and they are a child of God, but it's a little bit different when they are forty-five and have been squeezed by society's pliers. So the work with people is enhanced by an understanding of esoteric physiology. That's my goal. Why even do this, why even take the time? It's not about healing so much as it is about compassion. If I can be compassionate healing will follow from people's own insights.

I would like to quote from *Guidance in Esoteric Training:*

> [Feelings of anxiety] have no place today and will have still less in the future. What occurs when we feel anxious? The blood is driven back into the center of the human being and into the heart, to form a firm central point and make that human being strong in opposition to the outer world. (p. 111)

What a concept! What we are talking about is arterial circulation. My contact with the world starts with my lungs, where the venous blood is affected directly by the world. As blood makes its way to the center and a target organ, the experience we have in our soul is anxiety.

> ...the blood is driven back into the center of the human being, and into the heart, to form a firm central point and make the human being strong in opposition to the outer world. It is the inmost power of the "I" that does this. This power of the "I," which affects the blood, must become ever stronger and more conscious.... What is harmful and unnatural today, however, is the feeling of fear connected with this flow of the blood. In the future, this must no longer be so; only the power of the "I," without the fear, must be active....
>
> In order for human beings to become equal to facing the evil powers of the future, they must take hold of their inmost strength of their "I." They must be able consciously to regulate the blood in such a way that it makes them strong in the face of evil, but wholly without anxiety. They must have in their power the strength to direct the blood inward. (pp. 111–113)

This is a very big idea. Now the reason this is possible is that, in the arterial blood circulation, catabolism is the rule; it is the breaking down. You think, well wait a minute; that's the really good blood. It comes from the lung into the heart, the aorta, and then through the body. Everybody knows that's the good, red blood. Yes, that's the good blood, but it immediately contacts a cell, and the cell says: Here is my junk; give me your good stuff. It's true. The cells catabolize the red blood. So arterial blood is being broken down with every contact along the way. The cells want to unload their junk in exchange for the good stuff; it's a little like raising a teenager. Great dinner. I'm going out with my friends. Bye. Fifteen burgers, Mom. Thanks, bye. Give me the good stuff, here's my laundry, goodbye. You know that feeling if you have been a parent in that situation; there is a kind of anxiety that comes with that breaking-down process.

In your body this happens every time your arterial blood contacts every cell in your body, and every organ and every organ system. The arterial blood, though it is the source of nutrition and oxygen, gives oxygen to the cell and receives the carbonic acid of the cell's metabolic waste, and that extraction process and receiving the carbon is a kind of breakdown, because the substance falls into the blood and links with the salts in there, the phosphorus, calcium, and sodium salts, and creates the compounds that go through the lymph and so on up into the liver to be processed. With the emergence of the blood from the aorta in the heart, the oxygenated blood that has been detailed by the liver and filtered and made real nice after that—this is the beginning of breakdown in the body, even though it is also the emergence of this new potential. As soon as it emerges, there is a breakdown of not only substances that we've taken through our digestive track, but also as the arterial blood meets the outside through every glandular secretion.

When we have a sensory experience, a gland is affected and the nerve side of the gland causes a metabolic reaction in the vascular side of the gland. The gland responds to the sense impression in kind and puts a deposit in the blood to balance out the nervous energies stimulated by the sensory experience. The gland deposits in the blood adrenaline, nor-epinephrine, or whatever will balance the nervous impulse. It is the production of hormones and neurotransmitters that creates what we can call soul activity of the astral body, also known as the soul life. Sensory experience also affects the arterial blood supply, and substances continue to be deposited in it. The cells take in the oxygen and nutrition from the platelets and deposit their junk. So that quality—all those exchanges, the nerves firing, the pituitary secreting thyroid-stimulating hormone to regulate everything, and depositing substance all allow consciousness to arise. The substance has fallen from the levity forces of the life body, and we have an inner experience. We look through our eyes and experience something out there in our consciousness. I recapitulate the Fall. That's why I discussed old Saturn earlier, so that a light bulb would go off when we got to this. Sensory experience and the trains of energies linked to it are a recapitulation of the Fall. It is exactly what the Archai did that led to their fall, and it's happening in us continuously, and the constant falling of substances in us from the blood, consequently giving us both freedom in our consciousness and anxiety at the same time.

When you take a bite of your sandwich, your body does not welcome it; more likely, it says: What the heck is this? We have to destroy the sandwich because we cannot use it as it is. Just run a sandwich through the blender, inject it into your vein, and see what happens. Your body would go into shock because it is alien, even though you love it. Even chocolate is an alien substance, so when you take in something from the outside and put it in your organism, your organism resists it. It has to; your soul wants it, but your life body has to render it into a form in which the "I"-organization can dominate it. The sandwich has to be destroyed so that it can come back in the image of the great god: Me. It has to come back according to my needs for incarnation.

The arterial side takes the hit, which releases forces as a result of the depositing of substance from the blood, giving us consciousness. The life forces that were keeping all of the substances in solution in the blood are released back into a levity state when the substances get deposited. Just burn a stick of wood and you get a picture of the fire going up and the ash falling down. That release of the levity forces makes the levity forces available to the "I"-organization for consciousness; you know that the "I"-organization changes levity forces into thoughts. Here we get a picture of how the "I"-organization uses consciousness to build a structure, from its capacity to form thoughts from the released levity forces of the life body. Out of its spiritual background, the "I"-organization builds a vehicle, so to speak—architecture in another dimension that it can use for further experiences. That structure is called the astral body, or soul. The "I"-organization uses the levity forces to form thoughts and create a vehicle out there, or little model, and then it lives in that model, visits it during the day, takes a little vacation at night, and has the whole thing worked out over time. It becomes attached to the vehicle and says: That's me. But it's not actually the True Me, but the vehicle I use, and it's on loan. That's the reality.

This drama of the Fall is enacted in the arterial blood. By taking in substance, taking in sensory impressions, processing those sensory impressions and substances, creating secretions, excretions, deposits that go on then to build the sandbar of my physiology, my actual flesh, and the life forces that are received from the original hierarchical archetypes are released to the activity of the "I"-organization for the purpose of consciousness. The "I"-organization uses the life forces, then, to construct a vehicle, or model, and to live in it as a soul, and then the "I"-organization has domination over that soul, which, as we read previously, Steiner said that it was not the original picture but caused by Lucifer. There is a complication, in that the soul dominated by the "I"-being, then, has "educational" experiences, because the "I"-being has ideas that the soul doesn't get yet—such as being a really good person for eternity. But the soul insists it has to win. Win what? Don't you understand that you are already an eternal being? The soul wonders what's the problem with me living my life? However, to life a life, the soul says; give some of your forces to the life body; I need some of your forces. And the liver says, okay, I've got some life forces stored here. Physiology calls those stored life forces glycogen. The glycogen in the blood then goes through catabolism, breaks down, the water separates, the levity force in the water is freed up, and the carbon, hydrogen, and oxygen are metabolized.

This is the great cosmic drama of physiology. Under the influence of the "I"-organization, the astral body affects the life body and demands the forces needed to create consciousness. To do this, it must destroy something. What does it destroy? The "I"-organization, working through the astral body, destroys the life body. That's why I have to go to sleep and have to eat lunch. My consciousness destroys my life. When people hear or read that in Steiner's works, they think it's weird. And it is, but that is not Steiner's problem. It's our problem. Now what? As says in the Gospel: *metanoia*—change your thinking. Change the way you feel about things in your soul. Change your thoughts so that they don't take such a huge bite out of your life forces. Change the pictures you create inwardly so that they don't make such a demand on the levity forces released from your catabolic activities.

Because your "I"-organization monitors this whole process, your "I"-organization can manage to make these adjustments to the soul life. However, it has to be conscious of the activities in the soul to do that. Rudolf Steiner had a phrase for that: *consciousness soul*. My "I"-organization has the potential to use my vehicle in the right way in consciousness soul, because it understands the various levels of consciousness that make up my physiological profiles. Each organ or organ system is a different kind of consciousness, and my "I"-organization is a kind of sovereign to those various states of consciousness when it is a good ruler. But consciousness soul can also be an antisocial tyrant, and then the various states of consciousness appear as fragments of a True Self and they bring torches to storm the castle. If I'm a good ruler, the peasants are out working in the fields, singing songs, and being happy. But if in the consciousness soul level of awareness I am too tyrannical, autocratic, and egoistic, the fragments of my personality erupt into my consciousness at random carrying torches and pitchforks. This state of consciousness is illness.

The personality fragments are saying to the rulers that they need to change something in the way they rule—that it is not clear what they need to change because those personality fragments are carrying torches and pitchforks but don't know what is needed to change. The ruler needs to decide to change, because the ruler is the only one that can change.

The peasants—the elemental beings' contract says: You report to the ruler. They are not free to simply act. Their contract says, this is your game plan and they can't change just because their ruler is a bozo. They just have to do what they do: *heal*. On the healing side of all of this, after the carbon is deposited in the blood as carbonic acid, it starts to link with the salts in the blood, forming structures that will eventually move into the venous circulation to build new protein structures now that they've been broken down. The "I"-organization can enter, but in an unconscious way. If the "I"-organization were always conscious of all this, you'd be a very busy person. The venous circulation is the building side of the ether body, but it is unconscious.

The arterial circulation is the breakdown side of the life forces and engenders wakeful consciousness. The venous circulation is unconscious because a higher form of unconscious is known as *superconscious,* which, though unconscious, is the source of the creative imaginations that create the structures of the protein and all the chemistry and the source of the harmony of the spheres, the forms of the body, and the subtle play of molecular ionic attraction and repulsion. All of this is still in the hands of the hierarchies in our personal unconscious, the interaction of the life body and the feelings in the soul. These unconscious creative imaginations enter us with every sensory impression. They enter when we eat and when we sleep, and our spent life forces are renewed through them.

Think about it; if you had to do the work of this refurbishing at night, you would consciously have to organize all of that healing process. It would be worse than learning a new software program. The great gift of the hierarchies and the Godhead is that we don't have to do this. They do it for us. They fill our tank every night and then we go and destroy it, but they understand that we need to do that. They have compassion for us and fill the tank each night.

It is the same vehicle they refill every night, and until the one who is designing the vehicle understands that perhaps the vehicle needs to be designed in a different way by a change in thoughts and feelings, the good stuff goes in—new wine in old skins. Changing from the arterial blood to the venous blood is the end of the breakdown. The venous blood that returns ultimately to the liver includes the lymph and waste products of the whole lymphatic system. The lymph is the source of the immune system, so the cellular activity in the lymph structure monitors and tags the foreign proteins.

When we consider inflammation, we will look more closely at the amazing lymph—a whole world given to you by the hierarchies for protection. However, all that drama gets mixed with what you had for lunch yesterday and what you saw and heard. Those inputs turn into juice down in the digestive area, and from the capillaries on the periphery it unites with what flows from the digestive area making its way back to the liver. The liver takes everything that has been tagged and broken down and brought together by all these other systems and brings it all together into one place in such a remarkable and miraculous fashion that physiologists have no idea how it happens.

Totally contrasting chemical operations happen simultaneously in most of the micro-spaces of the liver cells, but they do not neutralize each other as new substances manifest. First it is over there, then goes the other way, and ends up at another place. It's like watching two flocks of birds silently and effortlessly separate from each other in mid-flight. No one knows just how all of these chemical miracles happen in the liver. Researchers can name the tissues and secretions using tracers to show where they go, but how they go is not understood because the action of the liver defies known chemical science. The liver is the source of life, the lifer—it helps us live, which is why it is called liver. It creates new life in us from what has been catabolized, or broken down, and then lifts it into a condition where it moves from dead material (an alchemist would call it the corpse) from something that has no potential for development.

Somehow the liver brings together all of these things that have been tagged into a meaningful wholeness. Then, suddenly, the dead fragments of life have living potential again. Nobody knows how it happens. It is the great resurrecting of the corpse, in the liver. We could call it "Golgotha." Without it, we would just have stuff in bags that tends to putrefy. If you could take all those fluids that flow through the portal vein and into the liver and put them into a jar for analysis, they would be just stuff. But when they come from the liver, they are living stuff. This is a big mystery. So the venous side is the circulation that leads to anabolism and building forces.

In the liver the levity forces come back into the substance, but this has to be done unconsciously so that the breakdown does not become involved with a building process. In practice, the breakdown and build-up happens simultaneously in all of the cells, but the net result is that the venous circulation gathers all that is broken down and delivers it to the liver. The liver ends the breakdown through sleeping and dreaming. The wise beings that govern the body give your consciousness some time away. You are off duty in your consciousness, while the hierarchies move in and get the liver to do what it did yesterday again today. This is the great enthusiasm for life that is the hallmark of a healthy liver.

A person with a liver problem loses enthusiasm for living. It is one of the key diagnostics. They don't want to get out of bed in the morning, or they can't get out of bed in the morning, or they don't wake up until one o'clock in the afternoon. They just want to remain unconscious. There are many reasons for this. It is the drama of the liver. Following the path of the blood, we have the new blood coming from the bones—the resurrection process from the death process. Then the spleen takes the old blood, separates the old stuff, makes a deposit of it, and offers new potential, though it is still not yet living. It is the spleen's job to separate out what is truly bad from what has potential to go elsewhere, sending it to the portal circulation.

How does something I do with eight-year-old kids cause sclerosis when they reach fifty? This is why I spoke of the Bach fugues.[*] Take the simple motif of fugue 1, multiply it through, do variations on a theme, and you end up with fugue 14; the whole thing has morphed as life morphs. Life takes something simple and reiterates it. The ether body is

---

[*] Visit https://steiner.presswarehouse.com/sites/steiner/Resrcs/User/BachFugues.mp3 to hear this talk as streaming or downloadable audio.

rhythm. Just keep morphing, says the ether body. The consciousness soul doesn't understand morphing. The ether body morphs to stay in balance. If I am a cerebral teacher and don't bring images, then some little kid who is eight years old and goes through eight years with Mr. Cerebral becomes rail-thin and pale with a persistent cough. Fifty years later, take that and multiply it with bad marriages, lost jobs, and all that we go through. The thought pattern says: If I don't have the right answer, Mr. Cerebral will give me a red mark in my book. Then, when I get home I'll get kicked, and I can't deal with that. It's not that we are evil, but we have to know what we are doing.

Rudolf Steiner speaks of this in *Balance of Teaching* and *The Foundations of Human Experience*. He says we should look at the complexion of the class. The kids who have a lung thing are not sitting in the front row but in the back of the class. They want some space between them and what is coming to them so they can work it out. Because they take everything into their lungs, they are melancholic. As Mr. Cerebral, I teach from my lung, and the child is a little "mini-me" Cerebral, receiving it in the lung. By the age of fifty, that child has deposits and possibly a precondition for sclerosis. Such people don't know why; "That's just who I am."

This is the mystery of how we do these things, but in the fugues I discussed, you take a little theme, tweak three notes at the end, and it becomes a whole new thing. Just turn the first phrase upside down, and hear what happened. This is how you get a person at fifty with all these various motifs, but with a certain tendency in an illness. If my thinking is too self-centered, it becomes difficult for my "I"-organization to bring warmth, because I lose my enthusiasm for being in the world; it is just a pain in the neck—ninety yards of barbed wire before breakfast. Why even bother? And that quality of consciousness creates increasing deposits of whatever. It could be lipids, calcium, sugars, fats, oils, or acids. It could be a lot of things, but whatever it is gets deposited to the point that the organism cannot reanimate it. The "I"-organization cannot penetrate the physical level and instead withdraws, leaving the physical level to dominate, and all the fluids fall.

That is Steiner's picture of physiology: the "I"-organization, astral body, and life body; this is the reason he goes through all of this. Many of these mysteries cannot be solved through biochemistry. Even continuing down into molecular structures leaves us far from the answers. Why is this? Consider the "I"-organization and how the Fall happened and its implications for the life body; because the life body is still living before the Fall. The life body is living before the spirits of darkness fell. The life body is still reflecting the imaginations of the cosmos. A physical body that is attached to it is hooked to that because, from those imaginations, we were supposed to have a physical body according to the big plan, but the physical body was not supposed to involve matter but be the activity of the senses—the forces and movements in the sense organs; that was the original plan for the physical body. When Rudolf Steiner speaks of "sense germs of warmth, enthusiasm, and will," that is what he is talking about. It was to be a physical body that would allow human beings to perceive the great wise movements of the cosmos and to celebrate them. That was our original destiny as human beings.

All hierarchies above the Archai, the spirits of darkness are reflecting. Kyriotetes are reflecting the Thrones; the Dynamis are reflecting that interaction. Then the Exusiai build a form of all that is being reflected. This sense of form is like the form of an idea that you might be struggling to understand. The idea has a form but no substance. To understand the idea you need to grasp its formal principle and duplicate that in your mind. As Thomas Aquinas put it, you equate the thing and the mind. The form that the Exusiai gave to the sensory organization was a psychic capacity to grasp the form of the idea received from the hierarchies above them. The Exusiai gave strength to the psychic capacity of forming an idea based on sensory experience. The form was then passed on to the spirits of personality. The possibility arose in them of a separation between the original unmanifested idea and the form of the idea reflected in the consciousness of the Archai. Instead of just reflecting what was presented, a little bit was pulled off that was not reflected back. This led to the experience of "otherness," which is non-reflective; it is self, the seed of self, or personality.

This is not substance. You would have to go through old Sun, where the Fall becomes amplified, to old Moon, where there is a rebellion of the luciferic beings, because they are at the human stage there. The Archai are human on old Saturn; the Archangels are human on old Sun; the Angels are human on old Moon; the retarded Angels fall when Lucifer is their guide. This creates Earth conditions, in which human beings are human on the Earth, but they inherit this Fall; the repercussion of the Fall is that the physical creates a very deep experience of separate self, caused by matter occupying the physical body through Lucifer's intrusion.

Were human beings always meant to be self-aware? Goethe said, "What would a God be who pushes things around only from outside?" The existence of freedom requires the possibility that a member of the hierarchies is free not to recognize the creation. We are the tenth hierarchy, whose task, although we are free not to recognize the creation, is to do it anyway. Steiner does not say categorically that it wasn't supposed to be this way, but that the wisdom of the Godhead saw that things were going to be problematic in the context of the original plan. However, in their great wisdom, the guiding spirits held back the original form of the human physical body in what Steiner calls the *phantom*. With the phantom safe in the higher worlds, they then allowed the Fall to happen so that freedom could be realized. They reserved the phantom as a contingency, because they knew that, if the Fall happened, there would also have to be redemption. They were wiser than we can ever imagine, because they held in abeyance the seed of the original form of the physical body so that freedom could enter, causing the Fall. Then, out of love, the task was given to a member of the Trinity to follow creation down and to suffer the pain of having a physical body that contained matter.

That phantom, according to esotericism, was what Christ released into the Earth on Holy Saturday. The Gospels say he "gave up the ghost." He gave up the phantom; he gave it to the Earth. It is called "the harrowing of Hell." When the earthquake happened—when he was put into the tomb—the esoteric understanding is that his physical body entered the Earth, but through death he had completely transformed the physical so that the substance was brought back into line with the original creation, because he provided the phantom as seed for the fallen state of substance. The matter in His physical body had been transformed by the deed on Golgotha, and then that seed was laid in the center of the Earth as a seed for the New Earth.

My understanding is that Christ went all the way—to the ninth level, the very core of evil. Christ went all the way to the deepest, most negative darkness of evil that could exist. The phantom body is in the ninth level of maximum darkness as a seed of light for the future development of the human light body; however, an understanding of this is available to us right now only at a very basic level, because we have not done the required work. Nonetheless, if we do the work, we will be able to do that at the Vulcan stage of consciousness. This is the picture behind attempts to understand physiology, because I have to bring picture consciousness to what has fallen into matter. I have to revive it with imagination and enthusiasm.

In daily life, we become separated when we wake up, open our eyes, and see a world out there. When I put something on my tongue and taste it, I get separated. Separation brings consciousness, and as a result I go through catabolism. When I open my eyes, I recapitulate the Fall of the spirits of darkness, because I open my eyes, and there is stuff out there. Then, however, I have to enter Archangel consciousness and I have to say that is out there because the Archangels received love from the Seraphim. I have to acknowledge that I must love what has fallen. Then I have to take the Angel path, because the Angels received the sense of harmony from the Cherubim. I have to say, not only do I have to love it, but I also have to harmonize my being with it. I have to have imaginations in me that participate in the becoming of the creation. Then I become the tenth hierarchy. This is one of the deepest mysteries in Anthroposophy. It is called the mystery of the phantom.

Chapter 5

Glands and the Heart

Let's break down some of the terms you can use to help us understand the discussion in this chapter.

Cosmic etheric: peripheral sculptural forces from starlight; these forces move from the periphery to the center in a planar movement.

Cosmic astral: motions of the planets that interact with the peripheral forces and convert them into etheric formative forces; these have a musical quality and give rise to elements and the forces and laws of nature. These have been astralized to move from a center to the periphery.

Consolidating forces: these take hold of etheric formative forces and the elemental patterns and convert them into substances, secretions, and excretions.

What forces constitute the cosmic etheric? Rudolf Steiner calls them peripheral, or planar, forces. Thus far, we have called them anabolic, venous, or unconscious. They are levity forces, cosmic ether forces that support the building process that leads to substance formation and growth. They provide a kind of energetic envelope for the Earth. Their source is starlight streaming through the Zodiac. They can be represented in the terms of physics as lightwaves and energy coming from all directions in space and moving into the Earth's realm. Esoterically, they are the energetic but invisible source of life on Earth, where they manifest as the energetic templates for substance formation and represent the creative ideas of the hierarchies that support manifest creation. These ideas are very fundamental to Steiner's imagination, and not only his imagination; they go back to the most ancient times as the source of life for humans, thought to come from the periphery, or stars.

The cosmic ether forces are planes of light, and physics recognizes that light moves in "wave planes" or "plane waves." The action of planes of light, from an esoteric point of view, is to pull consciousness toward the periphery. If you go out in the evening where the starry realms are clearly visible, such as in Canada or the mountains, you can see the stars out on the periphery. You wake up in the morning, and there's the Sun, which reveals what is on the periphery. The action of that light is to pull your soul out to the horizon, or periphery. Steiner calls that peripheral force "suctional." The action of the ethers, especially of the cosmic ethers, involves a dissolving, or sectional, force, removing substance from manifest conditions.

People say that the ether body makes you round. Is that the cosmic ether forces or the etheric formative forces? We need to make a distinction. Rudolf Steiner gives a picture of how, if we were in the realm of etheric forces as a human being, we would dissolve instantly and start flowing back toward the Sun. We would not have a body as we do now; it would turn into light and start radiating; we would be our own little sun, and we would return to the big Sun. The periphery is the source of abundant life, which is continually manifesting here on this plane where we are. When abundant life out there is present here, we call it death. I would have to die to be in that abundant life here, but then I'd be in abundant life at the periphery, not here. Here, we have life abundant "through the glass darkly," because what we call death is actually birth into eternal abundant forever-evolving life.

Through polarity, what we call living stuff that we can see or touch is actually a corpse of the true invisible life force. Thus, we have a learning curve esoterically when considering the etheric. When I look at my body, I say it is living, but if the life that animates it suddenly left, we would see reality as my body hit the floor. The part talking to you and the part hearing me communicate with you is accomplished via your vehicle, but it is not me talking or you hearing. The part that actually listens is you, but the vehicle that transmits the sensory activity is not you; it is only a "loaner," and after a certain age it is a beater that you return to the earth.

There is a quality of consciousness that is actually set against life. Previously, I linked this to the "I"-organization that consumes life force to create thought. When I awake in my body, I am annihilating life abundant by saying, this is me here, God. Pay attention, God; this is me. And God says, I'm busy.

From an esoteric perspective, cosmic etheric forces are actually dissolving what has become manifest. If we were in the cosmic etheric forces at this moment, our form would dissolve and we would die. However inside that distant realm of the cosmic etheric light of stars is the sphere of planetary movements. Movements of the planets in front of the planes of light of fixed stars create frequency modulations in the light and waves in the light that are not just planes coming in perpendicular to the line of lightwave travel. There are loops and all kinds of apparent retrograde motions that, as mobile lenses for the planets, create conditions that we here call form, or what Steiner calls etheric formative force.

Imagine that I have light streaming in and I take a lens and move it in front of a bunch of lights on a screen. I tweak it and rotate it as it moves in front of the screen. Suppose I have a whole bank of projectors with images on the back wall; I have a screen and a lens moving in front of it that I rotate as the lens gathers all that light and its images and patterns. I would see all of the pictures projected on the screen. Well, esoterically the pictures projected on the screen; those pictures or imaginations are the source of the organs and tissues of your body. The pictures out there are the ethers or the light-filled ideas of the divine hierarchies. These light-filled imaginations weave in space and bathe the earth in cosmic imaginations from all directions in space.

Imagine a weaving wall of light composed of the imaginations of higher beings coming from all directions. Then imagine that the loops and motions of the planets moving in front of the wall of light create the forms that manifest as the sense world. In this imagination, the lens moving in front of the light images is the astral realm, composed of the formative musical geometries linked to planetary motion. This musical field of form potentials creates form principles (Exusiai) that eventually become, in Earth's evolution, the form I call my arm, my heart, or my lung. I have a little astral realm in me as an image of that harmonic realm of weaving light. In that imagination, the Sun, when written small in me, is called my heart; the image of Jupiter, when it manifests in me, is called my liver. This is because the geometrical harmonic functions in the planetary realms manifest in me as the geometrical harmonic functions of how my astral body has configured the life forces in me to create an organ. As I create the organ, it seals off and holds in it part of the cosmic ether forces and borrows them, like Bach's second fugue (a dotted eighth note followed by a sixteenth), it borrows some of the energy from the great cosmos and locks it into the astralized organ. That is the action of starlight filtered through the planetary realm that supports the forces of life in my body.

The difficulty is that my "I"-organization, repeatedly pushing my astral body into my ether body over time, habituates some of the cosmic life forces in my life organ so they are no longer able to gain access to the dissolving forces from the cosmos. When this happens, the organ has become habitualized by the soul; the ether body becomes locked by continual astralization into a particular pattern. It is no longer recognized by the hierarchies, because its root is found only in my personality.

Because I am a human, archetypal patterns occur along these lines, some with my liver, my little Jupiter, and some with my heart, my little sun. We call these archetypes of human personality "temperamental dispositions" when in the life body and "characterological dispositions" when in the soul. When that happens we have the consolidating forces (as in fig. 2) as the lowest level of those forces; they take hold of the etheric formative forces. In a series of step-down transformations, the cosmic ether forces are changed into etheric formative forces by the cosmic astral in the harmony of the spheres. These forces then fall into consolidation as the elements. I share the elements of earth, water, air, and fire in my body with the forces of nature. In my body I have the same elemental forces that guide the natural world, which give me a magical ability. This is the worldview of alchemy whereby, when I think and do things, I create elemental beings from the forces that guide transformation in the earthly realm.

Human beings, as the tenth hierarchy, can do that kind of transformation of the elements. We call it architecture or medicine. But then I am linked to the elemental beings who serve my imaginations and "therein hangs the tale!" I am linked to the imprisonment of elemental beings in my elemental body through consolidation. Where do I get consolidation? Through my astral body, habitually thinking thought patterns that just keep repeating without creativity.

There is a kind of hierarchy that is the elemental world. The consolidating forces fall out of the ether forces and serve human beings in their illusion of a separate self, created by the manifestation of substances in the natural world. The physical body is the pattern of forces in my sense body, which carries the energy resulting from my sensations.

If I had a meter that could observe the electrical, ionic, or bioelectric flow from my eye through the nerve to my vision center in the back of my brain, I would see the quality of that flow as my personal etheric formative force that I pulled from the cosmic ether. I have made a form, a nerve, and I live in that nerve when I use my "I"-organization to form thoughts. Steiner calls this living in the nerve. The reason I can direct it through the nerve is that the force in the nerve would be there even if I did not have the substance of the nerve for transmission. It would still be there, but I would have the experience of what Steiner calls an aura, or the physical body without the substance. Phantom pain in a limb that is no longer manifest is just one example of this phenomenon.

In esoteric thought, the human phantom is the original aura. It is the original, and it was held off because the great gods saw the need for redemption as I said earlier. But the physical has fallen continuously into substance, and now we study this stuff; we look at the substance, parse it, and think we are just bags of minerals worth about $1.98 each at Walmart, a realistic estimate as mineral resources. But that bag of minerals is not you. The thing that animates the bag of minerals is your "I"-organization, using your astral body to deal with the ether forces to lift the minerals into life through the action of your liver, so that consciousness can arise. We need consciousness to arise, so that eventually we can redeem the elementals that have agreed to be a kind of sub-hierarchy for humans. On the next round, the animals will be humans, and we will be at the level of angelic consciousness. The elements we are creating will be somewhere in between, where we have to work it out with them.

Rudolf Steiner gives a very strange picture, which I have included here in a quotation about the fact that every device we make in the future round will be a demon.

> Our contemporary culture is itself creating horrifying monsters that will threaten the human being on Jupiter in the far future. You need only look at the huge machines that human technology is constructing so ingeniously today. Human being are creating demons for themselves that will rage against them in the future. Everything that we build today in the way of technical appliances and machines will assume life in the future and oppose humankind in terrible enmity. Everything created for mere utility to satisfy individual or collective egoism will be the enemy of future human beings. Today we are far too concerned with gaining useful advantage from what we do. If we really wish to help advance evolution, we should not be concerned with the usefulness of a thing but with whether it is beautiful and noble.... Everything beautiful and noble that we cultivate today leads to a strengthening of the good on Jupiter; everything that occurs as a result of egoism and utility leads to a strengthening of the bad.[*]

Steiner said this in 1908. We have locked into our devices elemental forces that will need redemption, because we lock in the elementals serving there, and they fall under the recognizance of the retarded princes of the earth. We lose contact with them, and the rule is, "What goes around, comes around." Eventually we are going to have to deal with the devices, because they will have the quality of consciousness. We are getting close to that in our present technologies. It will be another elemental consciousness. It will have to do with human beings allowing an elemental realm that divorces itself from the human to lift into a realm of consciousness.

---

[*] *Guidance in Esoteric Training*, Munich, Jan. 16, 1908, pp. 111–113.

That is happening. The elemental beings are not the creative intelligence that was known in the past as the gods. The elemental beings are the forces of nature who don't have imaginations of their own but obey the imaginations of the gods of nature—that is, until human beings force elemental beings to obey through technology. This transfer of control of the elemental world is the kernel of human free will. Human beings share control of the elemental world with the divine creative hierarchies and have cast down the spiritual beings that represent the forces of nature.

There is a process called the formation of the avatar, which we will discuss when we cover the descent of the heart. In short, the forming of the avatar is a process of the descent of smaller beings that are cast out of larger beings, which has happened since the very beginning. As a rule, when you create a being through descent, you become responsible for that being because, when you cast it off, it moves to a lower consciousness and you get a lifted consciousness. It is the same rule that guides the deposition of substance that allows expansion of consciousness. However the rule is that if you have the benefit of a lifted consciousness you have to take care of the being that was cast down as a sacrifice so you could be lifted. The Angels who evolved beyond human existence must now look after us human beings, who were cast down into human existence as the progressive Angels transcended. The evolved Angels now have to look after us, and the Angels that did not evolve are the ones who stick their foot out when we walk by, because that is now their job. They wanted to interact with their inner pictures in their own way, and that was the Fall into personality. They are stuck half way between Angel and human.

We will have machines that we will be able to bring close to human consciousness as part of the progressive forces of the cosmos. Other machines will rebel and refuse to be part of the progressive forces. It is just as the *Matrix* showed; that is what that film is about. I have some really great power tools for woodworking in my shed. Every time I put one away I say, thank you very much. You served me very well. In the agriculture course, some people commented to Rudolf Steiner about how terrible it is that people use tractors on the sacred land, and Steiner basically said they should get used to it because it will not go away. He followed that with a smile and said that our tractors will work better, if we make friends with them. That is very poignant. The consciousness we bring to devices we make in the future will be our ability to deal with the demons we are creating. It is not evil, but it is very unconscious; people don't realize that machines are in pursuit of being-ness as a new hierarchy, created and cast down by human beings.

Angels are not all good. It is merely a convenient New Age fable that, if it's an Angel or spirit, it is good. This is a fatal delusion for human beings. Some Angels are nasty buggers! They focus their whole being to induce human beings to do really stupid things. They don't *make* humans do stupid things, but they have the power to know exactly what bait to throw out to make you come out of your hole. That is what they understand, and they are allowed to do this because they serve higher beings who are also rebellious to the creation. Not all Angels are good. We have to be very clear about that. There are Angels who were retarded at the old Moon stage and whose purpose is to do evil—well, not necessarily evil, but *not good* and *not progressive*. My understanding is that the only ones who do evil are the human beings who know better but do it anyway. This is the work of black magicians. Not all Angels are progressive, and the question is: It is ten o'clock; do you know where the Angel is that you're dealing with? To say all Angels have good intentions is like saying all people have good intentions.

There is much that needs to be redeemed; redemption is endless. We have to redeem the chairs that we sit on; elementals are locked within the elements that make up the chair and in the forces formed by the geometry of the design, all sacrificing to be in that form so that you don't have to sit on the floor. Consider redeeming all the elementals that pass through our gasoline engines. Then consider all the elementals in the electric lights that we flip off and on. It all has to be redeemed.

Eventually we will be able to perform redemption of the elementals who serve the natural world. That will happen when we take responsibility for the beings who are serving us there. It is all about gratitude, love, and harmony. If you want to survive this, we need love and harmony. That is the way: Love = Seraphim; harmony = Cherubim.

Harmony is the basis of what I mentioned as *autonomy*. If I am given the ability as a human with an "I"-being to make decisions, my "I"-being is part of the I AM. This means that, if I am given autonomy and freedom to do this and can actually incorporate the activity of the I AM, I will do what is progressive for everyone, autonomously in freedom, not out of some sympathetic feeling that I have to blend with everyone. That is the highest moral deed that I can do; it is called love. This is "Philosophy of Freedom."

Steiner thought that human beings fail to understand this because we have lost sight of the spiritual world. Love and harmony that the Seraphim and Cherubim gave the Archangels and the Angels is our training for how to do this; it all has to do with pictures and images as the basis for the spiritual perception needed to develop living-picture imagination as part of the initiation process of mystery wisdom. We are being swamped by weird images so that we cannot tell if what is happening is true or not; the adversarial forces know that, if we learn to control the inner images in our souls, they won't be able to throw us their bait that causes us to fail; we won't fall for it. Imaginative capacity is just the tool for the later work on Inspiration and Intuition. But the development of imagination is the present stage of the great work. When I am acting truly autonomously from impulses of my "I"-being, I am acting in harmony with the whole. I am related to the whole, to everyone else. If my motives are twisted and based only on my personality, then I violate the principle of relatedness and fall out of harmony with the whole.

I use competency to check my motive for my autonomy so that I can bring my personality back into harmony with the progressive development of the whole. I can use that as a kind of failsafe, and then I find I am living in a world like Salieri in the Mozart film *Amadeus*. He is in an insane asylum, and all the people around him are doing crazy things, and he absolves them from their mediocrity, one and all.

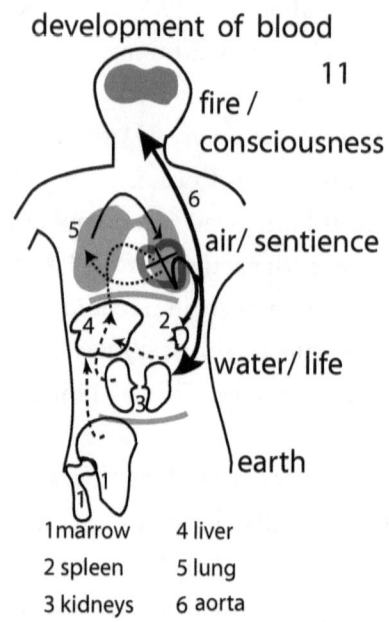

1 marrow
2 spleen
3 kidneys
4 liver
5 lung
6 aorta

That is the new initiate: I absolve thee, one and all. That attitude is necessary because we can't legislate against mediocrity but have to be able to surrender to the wisdom of something greater than we can understand without judging another being. If we can refrain from judging one another, we can still be discriminating about actions but not get caught by the adversaries. If we can just hold back from judgment, we begin to perceive the harmony of all things.

All of this has to do with the blood; my "I"-organization lives in harmony with the warmth condition of my blood. Warmth means enthusiasm to be, to do. Georg Kühlewind said, "Joy is moving the will." Enthusiasm is moving of the will—autonomy. When I feel I can actually make a decision and will follow it through, I go to any lengths to sacrifice time and energy when I feel I have autonomy in that realm. This is because I want to appear competent and have my actions be in harmony with the progressive forces in the cosmos. Why? My competency gives me a sense of relatedness with the best in the human being. It's obvious when we look at it that way. As soon as I have second thoughts or form judgments about the being of another, I can feel the shift in the energy. This has to do with how my blood is moving.

Figure 11 shows how we will work with the blood and how the blood passes through the body. Number 1, the marrow, is down in the pelvis and femur, and there is a flow from the pelvis and femur up toward the liver, number 4. That is one pathway. Then from the marrow comes the general flow of blood into the viscera, and that goes through the portal vein up into the liver, as well.

Glands and the Heart

The kidneys (no. 3) receive blood from the heart and return it to the heart. The spleen (no. 2) takes oxygenated blood from the abdominal aorta, does the separation, and sends it on to the portal vein of the liver. The kidneys (no. 3) filter twenty percent of the blood from the aorta, which then drains into the inferior vena cava, which drains the blood from the viscera. From there it goes to the right atrium of the heart. There is a whole flow of venous circulation going through the body to the liver to be lifted, put into the heart, and then into the lung. This also includes the fluids that comprise the lymph system and its function as the foundation of the immune system. Taken together, this is loosely called the *venous circulation*.

In these various organs, blood cells are modified to be red blood cells, white blood cells, killer T cells, or phagocytic cells. All these different cells are morphed blood cells, depending upon what organ they go to. The blood system produces multitudes of cells that eventually become the carriers of nutrition and a kind of failsafe in the organism.

Earlier I over simplified the pathway of the blood. The alpha for the venous blood is the liver, and the omega for the arterial blood is the spleen. I just linked them in an abstract shorthand of the circulation. In reality, blood comes from the liver and lungs to the heart. From there it flows to many different organs. The general circulation is from the heart through the arterial circulation to the organs to the capillary circulation to the venous circulation and then back to the heart. Blood also comes from the marrow and goes into the digestive area before flowing through the lymph and plasma up into the liver. If there were not multiple pathways of the blood, we would be a lot sicker than we are.

The great imagination of the hierarchies for the way the blood works in a kind of basic fugue-like theme (artery to capillary to vein) that keeps getting changed, moved, augmented, reversed, flipped, shortened, and quickened. Most organs have an arterial circulation that moves through capillaries to veins in a micro-arrangement of the larger circulation. Blood is miraculous. This is confusing if I expect the circulation to be like pushing pork into a kielbasa machine. This is not the way blood moves. There is a grand orchestration of fugue counterpoints. Our work with the imaginative exercise is fundamentally an abstraction. However, if we work with it meditatively, the abstraction of the circulating path of blood eventually leads to deeper and deeper insights about physiology. In reality, there are multiple ways to do this exercise. I will pick a general way through, but if you find that you have an issue with an organ, you can visualize the circulation in that organ as a useful meditation for insight into health issues.

Along these lines, there is a whole branch of anthroposophically inspired medicine called *chirophonetics,* in which doctors go over an organ while making sounds and gestures to clear it out. They don't touch you or send you to a pharmacy or whatever. What they find, especially with congenital conditions, is that this is very effective, because it works with a kind of soul property. The sounding and the movement together helps the body get the original form. Chirophonetics is the work of Dr. Alfred Bauer. There are a lot of modalities that come from such approaches that are not one-size-fits-all panaceas, but there are certain modalities for particular pathologies that are very effective. The use of eurythmy movements and sound over an organ is useful, as research has shown. A certain sound accompanied by a particular movement can connect a challenged organ to its archetype. Flower essences work similarly, as do homeopathic remedies. Forming inner pictures of sounds and movements can also be a path to healing. I can recall movements I make with my hands when modeling clay and, in my imagination, place them over an organ for the purpose of receiving insight about a pathology. Those imaginations become medicinal. Movements that I make with my hands—if I recall those feelings inwardly and make those movements and place them on an organ, I am lining that organ up with that archetype.

A sound, too, has a particular inner movement. Music has a geometric reality; it is more than just sound. Harmony of the spheres is the way form comes into being. I can make the forms in clay, remember how that felt in my hands, duplicate that inwardly as an imagination, and place that on an organ. It is like *T, S, R, M, A(h) – L, A(h), O, U, M* but more

Glands and the Heart

plastic in three dimensions. It is useful to use more than one modality of clay modeling and eurythmy as art experiences to experience a fuller range of possibilities for forming healing imaginations. What eurythmy and clay work have in common are imaginations.

The work with the blood here has many different ways of getting through the whole circulation, depending on where you are, how you are, and what you are working with. I've been working with this exercise for about a year or year and a half, and sometimes there is no way I can get down into my bones in the morning. I place my imagination on my thymus or my lung, and immediately I start yawning and tears come, and I have to leave it there for the day. If you get it right, your body will thank you for calling in and inquiring about its wellbeing.

So what we do with these exercises is just an archetype, but if we begin to educate ourselves about our pathology, whatever it is, we start to get a picture of the way it is moving. We can actually talk to our liver or blood and ask how it's doing and how we can help. And we will be told. Before I left the house today, I looked at my bag three times and thought I should check it. I looked in my bag and thought it was fine. Three times I looked. Later on, we had twenty minutes and were talking. I put my bag down and looked for my notes—no notes. I went out to my car and turned the key to come home and get my notes and something said, we tried to tell you three times and you were just doing your own thing. This stuff is coming in all the time, but we think it is something else because our thoughts are somewhere else.

So if we begin to practice and start body scanning, your liver, kidney, lung, or bladder is going to thank you for checking. Think about your bladder once in a while. You go there and then feel the knot in your back start to relax. If you keep doing this imagining as a practice, pictures suddenly come and you go to a website about renal dysfunction. You've never even heard the word, but it just keeps coming to you, and suddenly you realize that is the problem. This is how imagination works. But it doesn't work if you don't prime the pump with inner pictures. If you are priming the pump and doing the scan, you will be corrected. If not, it just

```
3 ┌─ ego organization- transforms growth forces into thinking
  │    source of healthy catabolism
  │
  │  ┌─ astral organization- personal soul forces and feelings
  │  │    source of the forms of the organs of the body
  │  │
  │  └─ personal ether forces - support growth of organisms
  │       source of anabolism
  │
  └── physical forces- mineral formations-elements
        source of forces and substances
```

keeps trying to put the message in the same box, and when the mailbox is full, it cannot receive more messages.

That voice mail is your "I"-organization in contact with your ego trying to download your destiny. Your "I"-organization is the part of your ego designated to monitor the way your physical body is interacting with your spirit. It interacts with your spirit through the interrelationship of your feeling life in your astral body, your soul, and your life forces. That is figure 3. There are two levels that could be described as "ego." One, with a small "e," the other with a capital "E." Capital "E" Ego, the "I," is a higher dimension that includes your whole spiritual life, not just the part of your "I"-consciousness connected to your physical body. Small "e" ego is linked to the Self we normally call "I," though we really mean "Me." I tend to use True Self for the capital "E" Ego, and ego with a small "e" for the lower self. The True Self, called the "I"-being by Rudolf Steiner, is experienced in deep meditation as the knower who knows it is knowing. This is a very elusive state of consciousness.

Using these distinctions, it is easier to understand the various levels of consciousness available to human beings. The True Self, or "I"-being, never incarnates in the physical body but uses the physical body as the prime source of learning in a particular incarnation. The ego is a shadow of "I"-being and is more of a personality than a true being.

The "I"-being, or True Self, experiences itself as a knowing agent in the act of knowing, while the ego experiences itself through memories based on past experiences. The "I"-organization is the part of the ego consciousness designated to monitor the way the physical body is operating under the impact of sensory impressions as it engages in the process of forming memories.

The "I"-organization is not the whole of one's ego consciousness, because most of our ego consciousness is focused on trying to fathom the "I"-being, which is embedded in the cosmic I AM (Christ). This is difficult if one's effort is based just on memories, because "I"-being is not the personality that forms memories. "I"-being, the True Self, incarnates only in high initiates. However, the ego comes close to incarnating through the formation of a personality based on memories, while one's "I"-organization is very involved in incarnation through the process of monitoring how memories are formed in response to sensory stimulation. The ego is a facet of the "I"-being, but it is not that higher self. The "I"-organization is the part of consciousness designated to monitor the physical body and form the ego experience. But the ego also has a whole other script that it reads from—that is, the action of the "I"-being, who you were last time and who you want to be at the end of time. That eternity is the time frame of the True Self. That "I"-being is looking at the I AM, the Christ, for guidance.

When I do that, the question is this: What does my soul have to do with my karma? What does the fact that I have this body, the sentient part of my body, have to do with what I am going through in my life? Those are big questions. And your ego learns from that and has a kind of designated offshore account into which it puts consciousness—a kind of tax shelter for what you did that you understand, of why you incarnated; about once every twenty years you get a picture of that. When that happens, you awaken to your "I"-being as a destiny parameter of your life that was previously unseen. In reality a part of your consciousness gets designated into that account of your "I"-being when you change a habit or make a resolve to do something better. Through such efforts, the "I"-being becomes more like the phantom, because it no longer has an agenda other than to be here now.

My blood carries the warmth or enthusiasm that my "I"-being is looking for in the world. I want to match my experience of the warmth in my blood with what I meet in the world. When I don't, I contract, and my blood goes to the center, because I feel like I will need to protect myself, and I begin to live in anxiety and aggression. This is fight, flight, or freeze mode in contemporary psychology. My blood goes to the center instead of expanding and being born into a higher state of warmth and enthusiasm for life. My "I"-being wants to match the warmth of the blood with the warmth in my life.

My "I"-being is my higher self waiting for my lower self to recognize that it *has* a higher self. My ego, or lower self, uses my "I"-organization to tell me whether I am doing that or not. When my lower self is discursive about my motives for doing something, my "I"-being approves. A little food pellet drops into the trough made of cosmic wisdom, and I grow in understanding—I expand my consciousness. But this expansion does not come at the expense of something related to drugs or "peak experiences." Cosmic consciousness is the expanded consciousness of the human phantom, whereby everybody is going to eventually get exactly what they need to evolve. That is sometimes called the Grail. Everyone will get what they need at the end of times. Everybody will be satisfied, and there will be a little piece left over with a bunch of beings that are satisfied with just being a pain in the neck. Rudolf Steiner calls those beings *Asuras,* another word for retarded *Archai,* the spirits of darkness, or spirits of personality, who became inflated by self-importance. When I become inflated with my own importance, it makes you less. When I make you less through self-inflation, I cast you down. That is the Fall. I do it by feeling separate from you, so when I cast you down, my consciousness feels a little bump up. This is another example of the Fall. The opposite, depression, happens when I feel that I am worse than I am, and in between is the Christ-being saying to just live and learn.

Glands and the Heart

The purpose of being here is to learn, not to be right, and that is the focus of the creation. I learn either by my willingness to learn—that is one school; or the other school is the cosmos with a big stick, making me learn. Leaning with the stick might be illness. When I inflate beyond my capacity for understanding, I draw the big stick of going to prison, being put into servitude, being shown that I am wrong, hitting the wall, getting sick, my house burns down, or whatever. Whatever your choice is, that is the perfect antidote. Steiner says you actually chose that particular obstacle in your last life to learn something critical to your mission as a human being.

Why is it so difficult to change my life, even when I really wish to do better? This is the great inflation; I think I'm better than I actually am, so I don't have to change that. But then when I do bad things, I know I have to change that, but the next day I forget. I am not alone in this, nor are you, and nor are the Asuras. They made a cosmic commitment to being pains in the neck. At the end of times, the Asuras will be a very small retarded hierarchy of beings that represent a seed of evil in the new cosmos. Without beings to pick up the mantle of evil, there will be no development, no movement. The way Steiner puts it—he radicalizes it—is that we should actually thank the evil ones, because they are sacrificing a vision of the Godhead to serve our development.

Now let's look at the following quotation.

> Just as the parallel organ to the ear is the organ of speech, so the parallel organ to the warmth of the heart is the pituitary gland, the hypophysis. The heart takes up the warmth from outside, as the ear does sound. Thereby, it perceives world warmth. The corresponding organ that we must have to produce warmth consciously is the pituitary gland in the head, which at the present time is only at the beginning of its development. Just as one perceives with the ear and produces with the larynx, so one takes up the warmth of the world in the heart and lets it stream out again through the pituitary gland in the brain. Once this capacity has been achieved, the heart will have become the organ it was intended to be.*

---

* Rudolf Steiner, *Foundations of Esotericism*, lect. 5, Sept. 30, 1905.

Why the pituitary? It is the place where the greatest Fall happens, because every sensory impulse that comes in, especially through the eyes, goes across the optic nerve and triggers the hypothalamus to trigger the pituitary gland to trigger the thyroid, stimulating hormones so that the whole body reacts to seeing the color green. It is a sensory stimulated organism, and I have to go into that process consciously to become aware of how the pituitary is responding to the sensory impressions that are triggering the glandular responses. I have to go into these processes consciously. That means I have to penetrate consciously the images that my senses are giving to me, to form an organ that will allow my heart to express warmth.

The pituitary is the place where Cain kills Abel, where the nervous system kills the vascular system, where the astral body destroys the life in the blood. A secretion that is created when the astral body destroys the life in the blood because of consciousness. That secretion is the hormones created by the pituitary as the master gland. Those secretions go out through the blood to regulate the whole inner life of the organism. My sensory experience triggers whole waves of inner responses in my pituitary that cause glands to secrete substances intended to balance the stimulus response patterns from the sensory experience. The glands are the seat of the ether body, because their function is to regulate the way the nerve and the blood interact.

This takes place according to the way the hierarchies have structured the body. My ether body is regulated by my glandular secretions, and the heart of the ether body is the pituitary. If I enter that heart, my pituitary gland, consciously, I start to experience how the forms of the world are creating in me inner responses. This is the role of the arts. I become sensitive to what I'm hearing, seeing, or touching, how I'm moving, and this creates sensitivity in me as to how my ether body, through glandular secretions, is changing the inner configuration of my astral body or how they are interacting with each other. It is in the place where my "I"-organization checks on how things are going. If I want to help my "I"-organization do that—and the "I"-organization does this by condensing

Glands and the Heart

life force into thoughts—I connect my thoughts to my sensory experiences in a different way. I have a blueprint for how to redeem the fallen quality of the substances that get deposited.

My pituitary gland represents the Jachim pillar that stood by the door into Solomon's temple. Jachim represents the process of how eternal life is turned into a corpse—how living forces in the natural world are turned into stuff. It is the symbol of life moving into death. I need to work consciously in that realm with imagery to feed forms to my body through the particular character of sensation in the arts. By sensing in a heightened way I bring a higher, more expanded consciousness to what I am hearing, and I connect that to thought patterns that allow me to redeem the fallen state of the unconsciousness normally present in sensory experiences. Normally, I am unaware of what's happening as the glands are lighting off in response to my sensory experiences. In the arts, I actually bring consciousness to that, and when I do I am a member of the tenth hierarchy and learning how to create a new world.

There is another pillar beside the door to the temple, the pineal. That pillar represents sleep, which the Romans called death's little brother. I have to take what I do in waking day consciousness—by training my pituitary and feeding it lawful forms—and take that process into sleep. That is the *Rückschau*.* When I take the images I structure into my ether body by paying attention consciously to the way my pituitary is firing off during the day, going back through them in the evening and taking them into sleep, I pass the pillar of Boaz through the door of the mystery temple. I go consciously through death to a new life.

Rudolf Steiner says that there are two pillars in the mystery drama, pillars on either side of the portal of initiation.** In my sensory experi-

---

* A preliminary exercise of pure observation and memory: "As the last activity each night, one goes over the events of the day, starting from the last thing one did and moving backward from that to the first thing in the morning. In principle, and after much practice, however, one need not stop there. One can continue back—over the previous days, months, and even years. In fact, there is no end to how far back one might go" (Rudolf Steiner, *Start Now! A Book of Soul and Spiritual Exercises*. p. 119).

** See Rudolf Steiner, *Four Mystery Dramas*, rev. ed.

ence I go from the eternal life that animates all of nature and all of the forms of nature into a death process that forms a corpse of the things I am looking at; I see it as a corpse rather than as a life experience. When I look at a tree out there, I say it is a tree. According to my sensory experience, it appears fixed. It is not fixed. But I can redeem the rigidity of my perception by understanding how the forces in the tree operate. I transform that perception into a magical formula through the arts. I paint the tree. I re-create the tree from my own forces, and then I have another creation to counter my imprisonment of the elementals in that thing. I heal that imprisonment.

If I take that healing imagery through the portal of sleep and dream between my pituitary and pineal glands, my soul goes into the spiritual world into a direct experience of my "I"-being, which says to my Angel: This guy is doing some interesting work down there and pretty soon may want a job here in the spirit as a member of the tenth hierarchy. It may be twenty-seven millennia, but the Angel is waiting for us to do that.

When we become the tenth hierarchy, the Angel can become an Archangel. In the next round, we will be the Angels, and the animals will have to carry the load, because they will have the consciousness we have now, but they are going to have to deal with the machine demons. We will have to be Angels and leave the animals in freedom and have the patience and compassion to monitor animals doing stupid things.

The big picture is that I have these forces of warmth in my blood that my heart is looking for in the world. And that is what I want to find, and when I don't find it I contract and have anxiety in the contraction because it means I have to prepare myself for battle. In the future when my pituitary gland gets organized, I will be able to allow the blood to come into the center without a worry, because I will have control over the image-forming process that causes my life body to go through an emotional cascade as a result of a hormonal cascade in my endocrine system. I will be able to monitor that because my astral body and life body will have shaken hands a little more because (one hopes) I will have redeemed my "I"-organization.

This is getting complicated, but I hope you get the idea. That quotation about the pituitary opens the door to why we are doing this meditative process. When I imagine my thighbone and the generation of blood in the marrow, I eventually begin to penetrate the forces of my life body, which are totally unconscious. Therefore, even when I get there I won't know I'm there. But if I keep going on the inside, I begin to awaken in a place where I am normally asleep. This awaking draws my astral body toward a liaison with my life body. In Rudolf Steiner's language, that liaison is called the sentient body, where I experience my sensations consciously; I experience what I am sensitive to. When I get near something and I have sensations, I can learn to bring them to consciousness. When that happens, my pituitary begins to approach my awake consciousness (normally it is not, because it's just part of my life body), but as it starts to come closer to my day-awake state, I become more aware of what I'm sensitive to. This is then our allergic response. The crazy thing is, as I become more developed, I become more sensitive and more allergic.

There is a kind of rule among physicians that a person who has many allergies seldom has cancer. Because of their allergies, they are constantly dissolving everything in the inflammatory process. When cancer starts to form, inflammation sends warmth into the thing, and it goes away. That is a rule of oncology. Cancer, the deposition or sclerosis side, is the opposite of the allergic side. However, if my allergies become chronic, my ether body eventually gets tired and just gives up. Then, I get cold and sclerosis begins. In general, we say that a person who has a lot of childhood allergies seldom develops cancer. But they can develop serious autoimmune problems that eat them from the inside.

In our exercise practice, if I start in my thighbone and imagine the blood coming out; I can then go in a number of different directions. We went to the spleen and the capillaries, and now we will add the kidney. We could say that one direction is from the spleen to the liver. From the pelvis to the digestive area to the capillaries to the liver is another direction. From the heart to the kidneys and back to the heart is a third circuit for blood circulation. Now we have three blood circuits we can follow inwardly. It does not particularly matter which one you take, but the one you take will have a lot to do with who you are. When I was working with this, I would try one circuit, and then I'd give that to my Angel. And then I would read something and find out...well, this other path really is the way the blood goes, and I had to ask, which is it? I began to realize that it is *all* of them. I began to recognize as I worked with this that, in a particular frame of mind, one of the circuits when followed inwardly would produce more yawns than another, so I thought maybe we ought to call this yawn therapy.

This practice is not foolproof, because it has to do with humans. However, there is a corrective in the life body. If in the process of producing thoughts my "I"-organization grabs onto an organ and stays obsessively in that thought pattern, that obsessive repetition of a thought produces disease. That is a problem. In reality, according to Rudolf Steiner, the "I"-organization is *supposed* to grab the organ, but it is designed to let go of it immediately. There should be an oscillation pattern of grab and release. The grabbing says to the life body and the physical body, I am here; it is catabolic. The letting go recognizes that we are all part of the great cosmos.

The goal of a healthy body is to have the "I"-organization touch in and then let go. That is exactly L, A(h), O, U, M – T, S, R, M, A(h). I touch in from the periphery and feel it (with the consonants), and I go out from a center with the vowels and let go. Then I go to the next organ. I touch in, massage it, pull back, let go. So it is not error-free, but the process we are trying to do is a fugue of catabolism and anabolism.

I started on the first day with catabolism–anabolism; that is the mantra. It is the picture of health; catabolism alternating with anabolism, catabolism alternating with anabolism. The healing remedy for catabolism, when things become too sclerotic, is L, A(h), O, U, M. The healing for too much ether force is T, S, R, M, A(h)—touching in from the periphery and organizing or harmonizing expansiveness through the consonantal forces of the ether body. The purpose of the exercise is to touch an organ, catabolize, anabolize, let go, and move on to the next organ and do the same thing, and move on through the body. Basically you are giving yourself a massage to check the organs, so that your "I"-organization can feel: I'm here and it's okay.

Over time you will have the experience that things are not working well. You will have a dream with a mood that starts to tell you about what is happening. Then the work begins. The purpose of this is to involve your imagination so that your dreaming begins to show *liver, liver, liver, liver.* Wait! I don't like liver. *We know; we think that you don't even like your own liver.* If I wake up every morning thinking liver, I would go and Google "liver."

What I am trying to share with you is that you can use this as a way of just moving through the whole organism. You can do this valuable work consciously. It is not foolproof, but there is a corrective. It took me a long while to figure this out so that we could do it. That is why I never wanted to do it with people, because the people may just want to concentrate on the liver for a couple of hours, which could lead to a gallbladder attack. You don't want to do that. A little goes a long way.

Now you can do the exercises. Go into the thighbone, then do the cycle L, A(h), O, U, M – T, S, R, M, A(h), and then let it breathe. We move through the organs. L, A(h), O, U, M is the healing sequence for sclerosis. T, S, R, M, A(h) is the healing sequence for inflammation. Sclerosis and inflammation are polarities. Sclerosis is the result of rampant catabolism; inflammation is the result of rampant anabolism.

You want to do both, because the "I"-organization needs to grab (catabolism) and let go (anabolism). Do both, and you don't obsess over it. Just touch it and let go. Go to the next organ, touch in and let go. When you do the cycle, do the catabolic–anabolic, let go and go to the next organ. Later when you penetrate what this kind of work can do for you, it is good to experiment with just doing the vowels alone or consonants. To begin, it is best to strike a balance.

Chapter 6
Nerve and Blood

We will now deal more with the esoteric side of physiology and less with the physical. I'd like to relate them to the forces in the soul, because the issue, as we heard from Rudolf Steiner, is the inability of my thinking in my "I"-organization to incarnate properly in my life organs. That is the fundamental problem of illness. One aspect of that work is to change my thinking, known in the ancient world as *metanoia*. But this is not the only thing we can do. Metanoia is the beginning. In changing my thinking, I actually have to enter consciously the way my feeling life is creating pictures in me, creating concepts from sensory experience.

When I have a sensory experience, certain feelings arise, but it is very difficult for me to contact them directly because they simply come up, and I recognize them only after the pictures are leading me off into fairy land. After I do something with my feelings, I think maybe I should have thought of that first. We all know that feeling. The soul is arranged in such a way that my thinking impulses are carried in my nerves, because my nerves are where the catabolic substance-separating forces occur. Steiner calls that a death process. My nerves participate in the realm of death so that I am free to change my mind. If my nerves had their own life, I would have to negotiate with them to change my thinking. But because they are dead, they do not add or subtract anything from my thinking impulse. Nerves simply transmit the sense world to me. I am then free to form concepts about what just happened in the sense world independent of the nerve with its sensory inputs. It is a kind of breathing process in the thinking.

The reason nerves are involved in it is because they simply transmit the two ways—from the concept to the percept as an inner path, and from the sensory world and its sense objects to an act of perception and then on to some kind of conceptualization. That is a second path involving the nerves and their function within the neurological sphere. We could say that neurology is the hallmark of the astral body. It produces day-awake consciousness. The astral body consists of three soul states—the *sentient soul,* the *mind soul,* and the *consciousness soul.* My consciousness is less awake in the sentient soul; I am more awake in the intellectual, or mind, soul; and I have great potential in the consciousness soul for superconsciousness, although that state is actually unconscious by today's standards. Superconsciousness in the consciousness soul, according to Rudolf Steiner, is incredibly antisocial.

To help us understand these categories, a picture would be useful. Those developments of sentient soul, intellectual soul, and consciousness soul are accompanied by the neurology of the autonomic nervous system. Sentient soul is the realm that governs my spinal column, cerebellum, and brainstem systems, integrating limb movements; intellectual soul is the realm of space and time, geometry, and planning of the limbic structures; and in my cortex I have the potential for consciousness soul, the thinking processes that lead to the capacity to be aware of my awareness. I can do that if I use my cortex to learn how to think without specific areas of my cortex. This sounds strange, but it is an actual neurological function—the default pathway system. That pathway is not through actual circuitry in the cortex, but is composed of random waves that propagate across the surfaces of neurological centers. Default pathway functioning happens when my more conscious "thinking" activity is silenced and a higher functioning energy takes over. That can happen only in my cortex, because there I can touch reason and morality. However, I cannot think morality as an assessment based on day-to-day thinking. That is the hard part. I can't really think morality; I can only *do* it and then think it after the fact. Rudolf Steiner says that I can think about the implication of a thought before

I have it and change it, but I cannot really know the implications of a moral act until I do it.

My moral sphere in reality is on the blood side of the equation, because the warmth in my blood is actually a gift from the cosmos to me that transcends my neurological functionality. Warmth in the blood is enthusiasm for being, the will to be, from the original deed of the Thrones. That is given to me in my blood, so my "I"-organization can learn to have the experience that it is membered into a community of spirits. Steiner calls this experience *ruling wisdom,* a term you find in the book of six lectures, *The World of the Senses and the World of the Spirit.* He talks about *ruling wisdom* and *ruling will.* The only time I participate in ruling will is when I go to sleep, and then I am embedded in ruling will because my awake consciousness of my "I" and the astral body tethered to it are off going to sleep school. In sleep school the "I" and the soul learn the moral implications of what was done during the previous day. You cannot do that in your day-awake consciousness, because you have the belief that you exist separate from all the stuff you are affecting by the things you do. You think that you have a kind of "spiritual Teflon," believing that what you put out does not come around. You believe that your anger is just righteousness and that your opinions are as they should be because there are bozos out there who need to get in line with your worldview.

Maybe I believe that I can drill as deep as I want for oil in the Gulf of Mexico because it says so in my contract. What is the morality of that? Unfortunately, as a human being I can check the morality of something only after I act on it. However, at night when I am asleep, the spiritual beings who provide the forces in my blood for warmth and light are active in my blood, creating images of moral possibilities. Those are the feelings of the possible morality that accompany the refreshment I have in the morning when I wake up. I am actually being given pictures of who I will be on Vulcan when I finally get my act together. In my normal consciousness, as soon as I wake up, they dim and then I begin to wonder how much money I will to make today. The polarity of the nerves and the blood create this kind of split in my soul between what I think and what I do. St. Paul said it: "I don't do the good I want to do, but instead do the evil that I don't want to do" (Rom. 7:19). That is it in a nutshell. That split in consciousness needs to be healed.

My birth splits the cosmos into the actual spiritual dimension of the cosmos and the stuff that fell out of that and is spread all around here. There is an original unity to these polarities, but my birth into a physical body splits them apart because of the way my consciousness is membered into my body, due to the Fall. Now, in between the nerve and the blood is my rhythmic system that allows my nerve and blood to interact. And the rhythm of my nerve and blood interacting, the pulsing in and the pulsing out, that is the touching in of the "I"-organization with its contraction and the release engendered as it touches out, the in and out of the rhythmic force is healing. The rhythm is the healing force between the nerve in which there is death and the blood which is eternal life, but to sustain life I have to have those two poles in my organism speak to each other. And if they do not speak to each other, I go through death into eternal life, but my life here I cannot sustain.

So there is a particular term that Rudolf Steiner uses to describe this rhythmic center, this feeling life in the center that unites the polarities. He calls it *Gemüt*,* an element in the soul whereby a person gets in touch with the levels of feeling that can simultaneously go toward thinking and the will. It is present in the animal forces in our organism as impulses for enthusiasm for the things we need in life, such as food; we could also call it drive or urge. Those drives and urges in us are actually a seed of what will eventually be our ability to do good; in the end days, I have to have as much enthusiasm for doing good as I do for today's lunch. The link is *Gemüt*. When I put the food out in the morning for Miss Kitty, her *Gemüt* goes, Wow—where were you? You're a half-hour late! That is enthusiasm for the food experience. And that is the seed in the soul of *Gemüt*, but Miss Kitty expects to download her food or she is unhappy.

---

*   *Gemüt* is variously defined as feeling, heart, soul, and mind. Linguistically, the term is considered obsolete—a pre-1901, Middle High German word.

As human beings, we expect people to do what we want them to do simply because we *want* them to do it. When they don't, we want to know why, and heads are going to roll. In this case, the will comes up to complete the thought, but there is no rhythm in the interaction; there is no flex in it. It's just lock, load, and boom, you're dead—I win...next. Whatever it is, whether it's my bank account growing or my thesis being accepted, whatever I want I am basically Miss Kitty waiting impatiently for breakfast. This expectation locks the nerve that is touching the blood into polarizing, because I cannot allow the deep, unconscious experience that I have at night when the spirits tell me what the reality of my life is—I cannot incarnate that lesson into my life, because my head is full of the way it needs to happen here in the land of stuff. I have no rhythm in it. Because I have no rhythm, I have to keep pressing it with my will, and it gets more and more polarized. Then I get hungrier and hungrier for control and for autonomy. I maybe try to fill the gap with competence, and then I'm misunderstood and things become even more polarized.

That is the *old knight* syndrome. I am working as hard as I can and they don't recognize me; I'm going to go eat worms and they are going to pay later. This is called aging. So you are on top of the heap, and then the young buck comes along with all the new information, and suddenly you are no longer the go-to person in your department anymore. This is the old knight syndrome. What you don't know is that you are actually receiving forces in your ether body every night that tells you how to deal with it. That is creativity in old age, or what Rudolf Steiner called *Gemütseele*, the enthusiasm for my life as it is soul and for what is actually happening in my life now; I can capitalize on it, because the forces are coming in every night in the dream. During the day, I oscillate between waking and dreaming, and when I dream during the day, if I have a way of gathering forces at night, those forces come out into the work I am doing and help it move forward. We could call it *grace*, because I don't do it myself; I'm not big enough.

Rudolf Steiner says that human beings cannot do what is moral; we need the gods to help us. To have the gods help us, we have to give to the gods the death forces that result in casting-down the creation that we do with our nerve process during the day. Then the thinking that is given to the gods as living, imaginative pictures is turned into will. It is a great help in this work that our movement across the threshold is a breathing rhythm between waking and sleeping and dreaming. Dreaming is breathing; what we don't realize is that, physiologically, we do it all the time. Every time we have a catabolic moment, we rise into consciousness; every time there is an anabolic moment, we go into the ether spheres, and the will in the blood comes forward to heal what we just catabolized. Hopefully, this language has some meaning for you at this point.

The development of *Gemüt*, the idea behind that, points toward a future condition when human beings will understand that morality in the world comes into them when they create the right vessel in themselves to receive the imaginations. The vessel is created by paying attention to the rhythm of going to sleep and waking up, because there is a kind of doorway for the transfer from here to the hierarchies and back again. The potential for this is physiological, but then the soul aspect of the nerve and the blood in rhythmic movement gives rise to the soul functions of thinking, feeling, and willing.

In thinking, in the nerve pole, my consciousness is dedicated to the past. In my thinking I can think only something I have already thought. It is impossible for me to think something that I have not thought before unless I establish a meditative practice to do that. That kind of practice engenders creativity. Creativity is a willed thought with no expectations of an outcome. The will in creative thought is the proper will; it is what we could call spiritual will, not the will that expects a result. When this kind of will dominates my thoughts, that is creativity. This is the "begging bowl" of Zen. My expectant will has to be empty, but my mind has to be present. In my inner work—especially in this realm of the *Gemüt*—as I start to do the work of penetrating my own physiology with consciousness, I become aware of a subtle level of rhythmic impulses of the way rhythm plays itself out in the world. Especially when I start taking things into sleep, I start to see that the waking–sleeping rhythm is a kind of archetype or analog of an immense number of things in the natural world. Huge numbers of things—climate for one; planetary motion for another; migrations of the lemmings for another, or whatever. As I start to do this work, anything that has a rhythmic nature to it starts to appear to me as a kind of waking dream. Steiner called this *inspiration*, but I have to structure the conditions before inspiration can appear. This is necessary so that inspiration does not become mysticism on the one side, or does not become what we could call sorcery.

Mysticism occurs when my perception of the world's rhythms allow my heart to go up into my head, thinking it's a beautiful world and everything is one with everything else, it's great and everything is being taken care of for me. That is my heart rising into my head and overtaking my ability to kill that kind of thinking. Why would I want to kill that kind of thought? Eternal Saturday on Maui. Rudolf Steiner calls it mysticism. As he puts it, when I don't carry my thinking process all the way to the end, I act impulsively and hope the Angels have a good safety net. That's the way not-for-profits sometimes do their fund-raising—with the attitude that the Angels will show up with the money when we need it. This is also symptomatic of the U.S. economy right now. There is a kind of mystical approach to fiscal responsibility. Voodoo economics, I think it's called. However, world economics is drifting toward the sorcery side. On the mystical side, I believe that if I don't do anything at all, it's all going to be great. This happens when my heart forces get too warm, go up into my head, and soften it. I start believing in all kinds of things that I cannot actually bring into cognition. That is the one side. Feelings move up into the thinking and dissolve it.

On the magic, or sorcery, side of things, when feelings move down into the will, the danger is that the will becomes a little hard and personal. If I begin to experience the rhythmic nature of my will consciously, I can learn that if I do certain things in a certain time, certain things will happen. I can see this in nature. The best place we find an analog of this is in how a predator analyzes the movements of the prey. The prey comes through here every day and moves this way or turns its head in a certain

Nerve and Blood

way when it eats. If I sit up in this tree and the wind is blowing in this direction, they walk under the tree and all I have to do is drop from the tree onto their back, and lunch is served. This is really the nature of the Ruling Will in the natural world. As a human being, if I observe that and get that consciousness, I try to structure my life so those kinds of opportunities for free meals happen. This is point-and-click trading strategies in the stock market. I can organize chaos to happen with a few clicks and then watch the feeling life of the other investors in that stock spin out. With my feeling will, I wait for the fear feelings of others to move their wills *en masse* in a certain direction. I feel that they will move the market to a certain turning point because of their fear, and I put that feeling-will into a computer as a stop or a buy signal. When the market trends to that feeling-will position, I will the money off of the table or put it on depending on how I feel things will go.

The technical esoteric name for this kind of feeling in the will sphere is called magic. We have equipment now to actualize huge magical will with numbers. With those numbers, I feel a weakness exists with which I can manipulate the feeling-will of others.

On the magic side, when I start to work that way, the temptation is to use my will in such a way rhythmically that I can affect something. On the mystical side, my heart, my rhythmic, goes up into my head and just says: You know, in the great rhythm of time it's all one anyway, and it's going to happen, so let's just not try to figure anything out and just surf the feeling wave of love.

On the will side, if I use this rhythm, I can perceive the rhythms of the feeling life of the thing I want and watch that. I can see just where that feeling is going to turn, and I put a little thing there. And I go over here and watch, and when it hits that thing, Bam! That's called a snare. Or a trout lure. What's hatching today? Oh, okay let me see, oh yeah I have one of those. So that kind of gesture in the natural world is perfectly natural. In the human realm that predator/prey relationship of feeling/will becomes magic. It becomes the ability of the will to create rhythms in such a way that I begin to control things.

Ancient culture was a will culture. Everyone except initiates was sleeping, and the techniques and the technologies and the operations were magical-will activities. If you wanted something done, you asked your astrologer, and the astrologer told you the time to do it. You did a ritual and repeated it at certain times so that you got the rhythm going, and it happened. You got what you wanted. The ancient world was a will culture; the people had not awakened to cognition in their astral bodies. They lived in the will with their feelings, and magic was the way things got done. Read *The Golden Bough* by Sir James Frazer. It is four-and-a-half inches of blow-your-mind book about magic as it was practiced in the ancient world. At first they worked with stones, then they worked with trees, and then they worked with people. By the time of the Renaissance, magic was about getting the blood of a baby and mixing it with the urine of a young boy and even darker activities.

The Jesuit protocols that Steiner talks about were given by Saint Ignatius Loyola and were known as devotion to the Sacred Heart of Jesus. Steiner felt that the protocols of the meditations did not allow those doing the meditations to do so in freedom. The imaginations were dictated to the practitioners and were rigidly adhered to in the Jesuit order. There is something a bit ominous about them, in that they seem to suggest something like a new world order based on the vision of St. Ignatius for the spread of his version of Catholicism as the basis for a world religion. Those meditations combine the mystical union insights of the saint with a magical-will impulse resembling a messianic cult. Steiner felt that was dangerous owing to misuse of the will.

Magic is the counterpoint to mysticism and has to do with the will linked to feeling. As human beings evolved their thinking, its process has become driven by the will, but jumps across the feeling life. This is because, through technology, people found that they can power the will with thought much better than they can with feelings because, if they get an error in machine design, they just change the error with their thinking and then make the machine do the corrected action. It's all about a better bulldozer. The goal is to sell a million of them, so thinking jumps the feelings, because the human *Gemüt* is where we need to make the great connection between thinking and the will.

Christ came and entered the Earth and said, "Do this in remembrance of me" (1 Cor. 11:24). Change your thinking first, but eventually you will have to go into your feeling space and clean out the closet. When you clean out the closet, we will trust you with the will to become a future world creator. Until you can clean out the closet, however, we are going to hold back the world-creating will from you, because when you lived in will before, look what you did. We know that was not working, and Christ had to come, because the world was being driven into darkness by the magicians. This was a very weird place back then. If someone did not like you, they would put your hair or fingernail clippings into a wax doll and throw it in a fire, and you could not do anything about it, unless you took a chicken over to somebody else. That person would say, "Oh, we can undo this."

Today we think this is all very quaint, and in a way it is, because we have stuff now that the old people could not even imagine in terms of magic. In the old days, they had to study for twenty years to gain clairvoyance. Today we just go down to Circuit City and swipe some plastic in a machine, and you have whatever image you want to see whenever you want to see it. It is called Google Earth. All you need is plastic. It used to be that you had to study with somebody for thirty years to get yourself clean enough to do what Google Earth does. Or clairaudience—the iPhone. It has been fixed into a device and we don't have to work the will to do it. The magic culture has provided the will for you through the creation of a parallel hierarchy of elemental beings known as machines. Machines are virtuosos of the will. You want me to do this six million times a day? No problem. I can out-will you, and now I can out-memory you a gazillion times over.

That is another hierarchy coming into being, using the forces of the Earth to pull this down into another realm of magic. I spent a couple of years on a nuclear submarine—that is big killer-type magic. Very unnatural, we could say. So that is a force in the world today down in the realm of magic. Then in humans, there is the equivalent to those machine forces in the mechanization of the soul life.

When I use my will unconsciously just to keep thinking obsessive thoughts about someone, through the blogosphere I can do damage to people in their feeling life. Machines let me amplify this unconscious repetition. What we don't really realize is that this same type of thing is happening even without the machine. If you have a habitual thought about someone—jealousy, anger, or whatever—you think they don't know that. In the realm of the will they don't, but in the realm of the feelings they start to pick it up. They just cannot bring it to day-waking consciousness, unless they start to do a practice. The downside of doing a practice is that you start to experience that level of relationship with people. You become more aware of how the power flows in relationships.

The organ that does that is the heart, the source of the *Gemüt;* the *Gemüt* looks for warmth of feeling. When it finds a relationship with a person, the heart may realize there is not a lot of warmth of feeling coming from it, though superficially there seems to be. There is a certain stage in development of this kind of consciousness at which such a lack in relationships becomes very evident. I have talked to a number of people who have done this work in Anthroposophy about this issue. The general experience is that, when they do this work and reach a certain stage where they hear the double-speak that people do, it drives them nuts, because eventually it's like they are opening a curtain and going into a funky little world where they can no longer believe anything. They hear what people say, but they also hear the actual astral intent behind what is being said. This is not pleasant and can be very confusing.

I have talked to a number of people who have studied Anthroposophy for a long while, and many of them have had this experience. It is a kind of splitting. This has to happen because you are crossing a kind of boundary where you have to realize that your thoughts have reality in the life of another human being, even if you never express any of it to that person. To do the work, you eventually have to experience how the toxicity of what you put out affects the people around you. This is rough, because it is ugly, and it leaves a bad taste. However, this is a certain stage in the path we could call *dealing with the proclivity to make curses*.

In the ancient world, cursing was just popping somebody's bubble again and again and again—similar to when a kid blows up frogs down by the creek. If you are a doctor, you probably blew up some frogs, or at least killed them in a biology lab to see what their guts look like. Why? Because you have to learn to deal with it. Thus, in this realm that you reach when you do this kind of work, when you actually go into your physiology and follow the blood circuits consciously, you can eventually feel and know how a particular organ reacts to what somebody says, especially in the morning. You wake up and there is a dream picture of that. If you can access that, you wonder how to deal with it and want it to go away. Next, you see in the other people how what you are saying will affect them; that is previews of kamaloca, little movie trailers of what you will see after your death. This is what you will do when you can no longer push people around or think ill of them. You will see and feel how they felt about you pushing them around.

Rudolf Steiner calls such training *Meeting the Guardian*. The Guardian tells you that everything that you think is coming at you is actually coming from you. What I am describing is called entry to the astral body. I enter my astral body consciously through the practice. The *Rückschau*, especially, does this, because when you go through your day backward, you winnow it, and you wake up the next morning your dream is right there where your ether body is affected. Your astral body is returning, and the ether body is assessing what happened when the astral body was away, but really you don't want to see that. Generally this is the cue for waking up. Those dream moods are symbolic of crossing of the threshold. That is crossing the threshold, but as I do this work, I began to cross the threshold *before* I cross the threshold. I start to wake up in parts of my astral body in which I am sleeping. And that is pretty much the sentient soul, in which I began to awake. As I do, I become aware of curses coming at me and curses coming from me. The teaching, at least in my life, is that the curses you are putting out are the source of the irritations coming to you. It is simply the great law and not personal, except that it is your own stuff.

At this level of work, the tendency is to foreswear magic. You had a certain thought with which to manipulate someone, or you had a burr under your saddle about a colleague, and you just keep saying the same thing over and over, and pretty soon you get dragged into this unpleasant place. This was called cursing in the ancient world. When you pull into the parking lot, and you wish that car was gone, that is a curse. If it happens every day, there is feedback that this is what you are doing morally. The Guardian is telling you that this is what's coming through.

If I could make little machines, put a curse into them, and just let them go and watch, it's a voodoo doll. Cursing is a dimension of health and illness. There is some part of my vital forces where my astral body—my inner life and my soul—meets my life forces that interface in the sentient body, which is aware of warmth or not from people. I navigate with that; I use this to choose friends and jobs, or not, or which industry to invest in. It is a kind of warmth organ in the heart; I feel an affinity or not. It could be ridiculously conscious; then it is magic.

Gray magic is cursing someone and white magic is praying for someone. It could be a sort of semiconscious repetition of an obsessive thought about someone, and we would call that grey magic. Rudolf Steiner talks about white magic in this context. White magic could be called healing and prayer. Steiner talks about white magic as things like the biodynamic work, about the need to have a different relationship to the rhythms of the way substances transform. This becomes alchemy, through which I consciously enter the elemental world and try to rectify the bad feelings, thoughts, and will impulses that I put out by creating healing substances and rituals. I do this by growing better food and by personally entering the elemental world working in the land where I live.

Or I might develop a sensitivity to how the rhythms of the planets operate in natural realms. The key to understanding this is, when I do these kinds of things, how I do exactly the same things that the black magicians do, but I do them with love and a prayerful intent. For most of the world, that concept is the hardest part to understand. This represents the fear of esoteric practice and even a large part of the population being suspicious of meditation. However, as Steiner puts it, the techniques of the left-hand, darker path and the right-hand, light path are exactly the same. The only difference is the intent of the operator, which brings us back to thinking, feeling, and willing.

What Steiner did with the therapeutic work—he calls it "the new art"—is the same as when the people used to intone into smoke as the Egyptian priest did. Steiner calls the future healing modality based on artistic perception "singing into smoke." According to him, it will be a recapitulation of the toning that ancient Egyptian priests did into incense they burned to make beings appear. This is the next art. Eurythmists all think eurythmy is the last or final art form for human evolution; eurythmy is premed school. It is not generally realized that when a person learns to control bodily movements or practices conscious speech formation they are forming an instrument for magical practice. That is a sobering thought—or it should be!

In healing and in illness, this dimension of cursing—what we could call cursing or putting out repetitive thoughts with a kind of perverse will, anger, jealousy, or whatever—the amount of energy involved psychically behind the thought also has a vector quality in the effect of the intention. If I am insanely jealous, I have much more intent in my willed thoughts than I do if I am just trying to work up to some jealousy. This might end in road rage, or the cops come in response to domestic abuse, and so on. From an esoteric point of view, there are actual beings that come to the soul of the jealous person in that situation. Those beings are attracted by the loss of psychic integrity in a person who is out of control, and they feed, so to speak, on the psychic energy of loosened consciousness. In the ancient world, the beings who visit an unstable person would be called demons; in today's world the diagnosis would be that the person has some syndrome. Those syndrome–demons become allies in doing things to others to get what you want.

Esoteric Physiology

I have learned in my life that, as you do the work and become more sensitive, you become aware of the energy around certain people. Disturbing energies emerge, but because of an established practice it is possible to become much more aware of what happens when things are said or done that create soul imbalances. It is a subtle realm. The more you do the inner meditative work, the more you are prone to infection from the beings created by people who are jealous, angry, or manipulative. You become more vulnerable to it because you are more open. Many people say that this extreme openness to spiritual beings is what eventually killed Rudolf Steiner. There were brotherhoods of people who knew how to generate beings to do dark things to people they didn't like. Madam Blavatski was put into a psychic prison by members of a dark brotherhood; she would try to think something, and her thoughts could not penetrate out or in. She was locked into repeating her own thought patterns. She was imprisoned psychically because she had the potential to reveal certain things that the brotherhoods did not want revealed. I've been told that Ehrenfried Pfeiffer said those brotherhoods were upset because Steiner was revealing the secrets of the heart. He was set upon by brotherhoods with magical practices to curse him. If you want to comprehend esoteric physiology, these things are at the root of understanding health and illness and balancing oneself.

Three years ago, my father passed away because of diabetes complications due to gangrene. He was the firstborn in his family, and I was the firstborn in my family, and there is a kind of passing of the mantle of dysfunction through such relationships. In the process of his passing, I contracted the necrotizing bacteria that was causing his gangrene. I got it in my blood, although I don't have diabetes. I kept getting infections that were difficult to treat, so I went to the HMO. I did all the blood tests and got the lab report, and they said I was as healthy as a horse, except for that rare, nasty bacteria in my blood. When I researched the bacteria, I learned that it is found only in the leg lesions of a bedridden diabetic. I talked to a doctor about that, and he said it was impossible for me to have it in my blood, even though the lab report said I did.

It was a kind of wake-up call that certain of my intimations were real. At the time, I was struggling with what we could call the family elemental spirit. I felt that a being was infecting all the men with the diabetic tendency—according to Rudolf Steiner, the inability of the "I"-organization to live in the body's warmth. The "I"-organization cannot penetrate the body, so the body tries to compensate by making layers and layers of protection on the outside as it corrodes from the inside. That is the pattern.

Extending Practical Medicine has a whole chapter on diabetes.[*] Because it was a scourge in my family, I wanted to understand it. Diabetes took my father's life through gangrene, because his ether body could no longer support his life functions with levity forces. His true Self had abandoned his life mission, and the life body experienced sugar as a poison. Sugar created a condition so that it became a ferment; his blood had become a ferment for bacteria. The necrotizing bacteria is one of those special hospital guys that come in, and I contracted it in the last few days of his life by just being there with him. I started having all sorts of spontaneous infections.

I wondered what was going on and started to work with the elemental being of the family. I began to understand that subtle realm of how our attitudes toward one another as the men in the family were creating a pre-condition that did not allow the "I" to incarnate fully through the "I"-organization. We were blocking the activity of the "I"-organization by ridiculing each other. It was a kind of a family sport that created permanent soul cramps in the men, causing them to feel like they were living a senseless life. Among the men, if you were not a bully you were ridiculed. If you were a woman, you were seen as just a servant. But if you were a young male, they saw you as a competitor, and the only way to deal with a competitor was through ridicule. This tended to destroy his self-image and hamper the ability of his "I" to tolerate being in his body. I looked at the cultures in which diabetes is rampant and found that this is often the way people treat one another. This is machismo, and diabetes is rampant in many Latin cultures. That is the way the

---

[*] Chapter 8, "Activities in the Human Organism: Diabetes Mellitus," pp. 42–46.

men treat each other, and my culture was the Polish coal-miner version of macho.

I started to become aware that there was this connection between what I later learned was cursing and this attitude of feelings that were repeatedly projected. Repetitive cursing was a kind of shock that drives the "I" out of the body by generating feelings where one's Self is no longer comfortable tolerating the circumstances. In such situations, the "I"-organization had to check out, leaving the body without support. When the "I" checked out, it became possible for pathogens to invade the life body, because the ether body was not being engaged properly by the "I." It was left to its own devices, allowing rampant inflammation.

Over time, the infections became so numerous that I was taking antibiotics, the generally accepted solution—let's just kill it all and start over again. I started to have very bad dreams, very violent, scary dreams, waking up and feeling that someone was in my room and attacking me. I had graduated to this level of experience. I didn't know what it was, but it was somehow a reality, and I was unable to deal with it. In the end, I met a doctor, a very gifted healer, in Colorado. He is an amazing man, a very advanced person. He has ways of analyzing the way your bodily sheaths are working, and he asked me if I knew someone who really hates me. I said, well some people don't like me, but I don't think somebody hates me. And he said, well, give me some names, and I started thinking I can always come up with a few names of people who don't think I am the cherry on the sundae. They would prefer that my car is not in the parking lot. So I hit a particular name of someone with whom I had struggled in the past. The doctor said that this was the one cursing me.

Suddenly, I realized something on a whole new level. I understood the reality of the relationship among the soul, the spirit, and the body. I started going backward, thinking through the circumstances of the relationship I had with that person. I recognized a certain feeling that I can now identify when that type of relationship appears in my life. If you and I sat and talked about it, and you could do this, you could eventually reach a place where you could identify that feeling, and then you could do things when that feeling happens to protect yourself from the unconscious, or even conscious, dark intent of others. Basically, it is the same as what doctors do to you, except they try to do it with substances only. Along with substances, this healer in Aspen gave me a verse he had received years earlier from a healer who worked in the Kabbalah tradition.

I will describe this in such a way that it relates to the Bach fugue.[*] What Bach did with the fugue, we can do with our soul force to create a fugue-like verse of repetitive thinking, feeling, and willing.

> By the power of the Christ all curses,
> whether ancient, hidden, or repeated,
> are found, bound, and defeated
> and must leave me now.

Then there is a series of three "to be" statements:

> To be revealed by the light
> To be absolved by the truth
> To be returned to the source of all things

This is a verse, and I think there are nine phrases in it, each with three parts. In each level of the verse, there is thinking, feeling, willing – thinking, feeling, willing—thinking, feeling, willing. The hard part of doing these things is that you don't want to struggle against "the onslaught," or attack, and you don't want to use negatives. Working with someone who consciously attacks you psychically is a very delicate matter. You don't want to curse a curse, because that just compounds the chaos. It is hard work to find things that allow us just to posit the integrity of our true Self.

Over time, what happened in my life as I repeated this verse again and again was that I realized the work I was doing was within the lawfulness of the whole, because I was not trying to cast anyone down. When I

---

[*] Listen to his lecture on a Bach fugue, available as a streaming audio online (https://steiner.presswarehouse.com/sites/steiner/Resrcs/User/BachFugues.mp3).

realized that, I felt free of the cursing will of that other person, and this was when I started to heal. I had mononucleosis recently for six months, and that was the final surge that started with the infection I got from my father on his deathbed and continued through the ill intent of someone who was practicing magic against me. The series of infections finally settled in my throat, because jealousy over the work I do was a big part of the curse. That was the battle going on in my throat. I was doing kung fu with some magic guy.

Now I will explain the structure of this verse, line by line, and finish with just the verse itself.

> By the power of Christ, all curses,
> whether ancient, hidden, or repeated...

The first line starts the pattern of thinking, feeling, and willing. *Ancient* refers to thinking, which is always in the past. *Hidden* is the undercurrent of feelings constantly being dreamed unconsciously in the soul life. *Repeated* refers to the will. If I repeat a curse, it has to be an act of will rather than just a random event. If it is an ancient curse, it is often a family or karmic thing. If it is a hidden thing, someone is trying to make juju on you in a secret way; it will affect you in your feelings as unconscious anxiety. When a curse is repeated, someone is actually after you and laying something on you as a magical act.

The line continues:

> ...all curses,
> whether ancient, hidden, or repeated
> are found, bound, and defeated.

They are found by coming into my perceptive consciousness (thinking) When I can think them, they are found, and then in the present I assimilate their power because now I know they are there. In this way, I block them from accessing me unconsciously. To the bad intent I am saying: I will tie you up and leave you here in the present; feelings connect me to the present. Then, after they are found and bound, they are defeated.

This means it is for the future, so I no longer have to deal with them; they are defeated and told to leave. This activates the will.

The first line is "By the power of the living Christ, all curses, whether ancient, hidden, or repeated, are found, bound, and defeated and must leave me now." This is the invocation and sets the stage for the rest of the mantra, or invocation:

> By the power of the Christ all curses,
> Whether ancient, hidden or repeated
> Are found, bound and defeated
> And must leave me now....
>
> To be revealed by the light
> To be absolved by the truth
> To be returned to the source of all things.

To be revealed by the light of thinking. To be absolved (dissolved and forgiven) by the warmth of compassionate feelings. To be returned (willed back) to the source of all things. These three continue the theme of a repetition of thinking, feeling, and willing as a fugue-like motif of this mantra's healing force.

Then there are three statements of the time frame covered by this mantra: "From this time forward!" This asserts that the past is no longer part of this mantra. Thinking in the past tense is finished. "Right now!" This brings the action into the present. Feelings exist in the present moment. "Forever!" This puts the action of the mantra into the future, the realm of the will. Then as a kind of seal, "So Be It." This is the assertion that removes the curse from the past.

"It Is So" is the assertion that states that it is happening in the present. "Let It Be Done" is the assertion that pushes the mantra into the future. Then the big ticket: "Thank You," the expression of gratitude. Here is the whole mantra:

> By the power of the Living Christ all curses,
> Whether ancient, hidden or repeated,
> Are found, bound and defeated

> And must leave me now
> To be revealed by the light
> To be absolved by the truth
> To be returned to the Source of all things
> From this time forward, right now, forever.
> So be it, It is so, Let it be done.
> Thank You.

Very clean, with a lot of integrity. I can tell you that it works.

In an activity such as this, the inner activity and feelings that surround this are very useful as you begin to open up to spiritual experience. As you do this work and begin to enter your own body, you start to open up. You become more aware of the subtle realms. In reality, we are always aware of the subtle realms, but we have layered over them with all kinds of beliefs and fears, and it becomes difficult to get a sense of healing. I have found that, while using this mantra when I was really sick, the worst part about the sickness was feeling helpless and afraid. By working with this verse, I found a way to feel I could stabilize my inner life without creating more turbulence. This allowed me to feel I could do something even though I was not well.

Once I realized that someone was actually sending me bad intent, this mantra allowed me to face that person inwardly without blame and ask: Why are you doing this? When I could ask this without anger or thoughts of revenge, I could pull away from the stickiness of the relationship. Once that happened, I could actually start praying authentically for that person. When I finally realized that the person was doing something dark, and because I was trying to meet that with something standing differently in the world, I could actually say prayers for that individual. I could experience that someone was doing something and was probably unaware of the implications that would need to be faced in the future. I was glad to be pulling away from that energy field without struggling against it. It was a release and I started to get better.

I had revised the verse from what I received from the healer. As I worked with it, Steiner's view of thinking, feeling, willing became apparent in the moods of the rhythm. I recognized a fugue in it. This was around this time when a message kept coming to me: Listen to the Art of the Fugue. When I started listening to the music, it became very clear to me that this mantra was a fugue with the theme of thinking, feeling, and willing, so I just added things and changed a few things around, and I asked that doctor for permission to do that. He said it wasn't his, so feel free. He said that people don't like the word *curse*. They think it is spooky, and in a way it is. Then he asked me to write a short article about the mantra and my experience with it, which he would then give to his patients to explain it, to say this is what it is. I decided to do a lecture instead and send him that.

Chapter 7
Esoteric Embryo

To create a kind of vessel for our work, we begin here with the sequences of the vowels and the consonants. They are not mine, but came from Rudolf Steiner. He gave them to the therapeutic community of doctors and eurythmists as archetypal soul movements to deal with the two polarities, sclerosis and inflammation.

In Steiner's scheme, vowels and consonants represent the archetypical word. In your inner work with these two sequences, you can experience certain fundamentals of physiology. The consonants, according to Steiner's view, represent periphery forces of the etheric, or starry, realm. They are the life-giving, regenerative forces. We could call them original forces. But the forces of life working at the periphery in the etheric realm dissolve things. This can be observed in the elemental world when we consider how water, the carrier of life, is the universal solvent. In physiology, liquid, or fluid, is the carrier of life, and Rudolf Steiner understood that fluid nature, what we called the ether nature, originated in the farthest realms—in starlight—in unmanifest but highly organized energies.

In *Extending Practical Medicine,* I read about plant sap and the existence of sugar, and that hydrogen–carbon–oxygen exists as a potential not yet manifested when it is in the fluid sap, because it's engaging the life of the plant. The image of the notes in a melody that has stopped playing is sugar. Those notes are consonants on the periphery; those notes, hydrogen–carbon–oxygen, are seen as cosmic realities. *Cosmic* is a code word that simply means the potential state. The cosmic state of carbon is its vast bonding potential to become many things. Carbon as a substance that appears bonded to something else has fallen from the cosmos; it has lost its potential to become anything. This manifest state appears as the substance carbon bound to water in sugar. Thus, carbon, hydrogen, oxygen, and nitrogen are the notes of what will be the melody of the life patterns of an organism, but these cosmic notes will not be the manifest part of the organism but the processes that carbohydrates and proteins engender in the organism. The cosmic form of carbon and the others represent the activities in the organism. This is what the consonants represent—the periphery forces, from Rudolf Steiner's point of view. You can read about it more specifically in the therapeutic eurythmy course.*

The sequence *T, S, R, M, A(h)* was given by Steiner as a consonantal healing sequence for when the astral body is out of control, to bring the inflammation into contact with an archetypal etheric form. *T, S, R, M, A(h)*, was seen by him to be a form—he called it a sheath—that surrounds the astral body when it is a little out of control. This would deal with an out-of-control inflammatory response. When the "I"-organization is present, an inflammation will automatically lead to a dampening down. This is called the immune system. However, if the inflammation is chronic or repeated through a person's attitudes, eating habits, feelings of self-worth, and so on, the soul condition—inflammation—becomes chronic and remains inflamed. In this case, the etheric body cannot bring the inflammation back into an anabolic state, or building-up condition. It is breaking down all the time.

Rudolf Steiner gave *T, S, R, M, A(h)* as a sequence for this condition. Traditionally, it has been used for hay fever. This was its original purpose, but he said that it is good for all kinds of allergic and inflammatory responses that cannot be controlled by the "I"-organization. It surrounds the inflammation with calm to create build-up, since inflammation, in the beginning, goes through a serious breakdown and then has to build up. When it cannot do this, you use the sequence *T, S, R, M, A(h)* for hay fever. Because of an allergen, your response is way off the chart and the sequence helps hold your ether body closer to the astral body. It tells your astral body it's okay, someone way out on the

---

* See Rudolf Steiner, *Eurythmy Therapy*; also, H-B. von Laue and E. E. von Laue, *The Physiology of Eurythmy Therapy*.

periphery loves you. It's all right, you are upset, we understand, but you don't have to be. That is *T, S, R, M, A(h)*.

However, it is not *T(ee)*, but *T(uh): T(uh), (e)S, (e)R, (e)M, A(h)*. Working with the *T, S, R*, I experience the three planes in space, the first gesture of breaking the egg into an organism. With *T(uh)*, you move up and down simultaneously; with *(e)S*, side to side; and with *(e)R*, front to back. I have had eurythmists tell me than no one ever said that. That's true, but this is what I perceive. If you follow the movements, you are moving in those three planes with those three consonants; the *(e)M* is meant to be balancing; the *A(h)* releases and reopens you to the periphery, allowing the healing force to enter. It is an anabolic sequence through the consonants. It is not mine; Steiner gave it to the world as a kind of archetype of a healthy ether organization. It connects with the three planes in space and the archetype on the periphery, balancing, releasing, and harmonizing: *T(uh), (e)S, (e)R, (e)M, A(h)*.

Then the opposite, *L, A(h), O, U, M*. Steiner gives the vowels as movements and, at the end of an anabolic phase when there is a build-up, there is always a breakdown; this is the action of the astral body. In the breakdown, things continuously break down and eventually form stuff. That is part of the normal sensory activity. When you have a sensory experience, your astral body starts to break everything down and makes secretions. Every act of sensation, whether between a cell and the blood or between you and your neighbor, results in some kind of deposit and secretion. That is the action of the astral body taking life from the life body and giving off what is left over. Rudolf Steiner calls it depositing, secreting, or excreting. It could also be called *hardening*; if I have a pea and put it in a balloon, I'm okay, but if I keep putting peas into it, eventually the balloon becomes hard; that's a tumor or rampant sclerosis.

Sclerosis occurs when there is continuous breakdown and things are hardened. When the astral body reaches the limits of its breakdown and can no longer move and there's all this stuff, I do the sequence of vowels, *L, A(h), O, U, M*—the movements of the planets in front of the fixed stars. It is the vowel movements that bring the astral body back into harmony and eventually soften the sclerosis. We lift it back into anabolism, because the purpose of the astral body is to break down and then build up. The purpose of the ether body is to build up so there can be a breakdown. Those are the two poles. L, A(h), O, U, M is a vowel sequence for dealing with sclerosis, catabolism, breakdowns, and deposits. Steiner recommended it for cancer and any disease that involves a deposit, such as gout and diabetes. Some say that these involve inflammatory conditions. They do, but they begin as a deposition or sclerotic gesture. In diabetes, sugars in the blood cannot be metabolized or brought into contact with the forces of levity. That is a deposition.

There is a key to this, and Rudolf Steiner gives the picture that there's no such thing as a sick ether body. The ether body is a continuous source of health, but the astral body takes forces from it to create consciousness, and illness comes out of consciousness When the astral body, or soul or feelings of my life, interacts with my life forces, the result is a catabolic breakdown. If my sensory life and my thought life are so intense that they dominate my organism without any refreshment, resting, or contact with the building-up forces, then I get diseases characterized by deposits. It could be sugar, acids, bones, fat, or endometriosis as connective tissue. Connective tissue is great and we need it, but we don't need it all over a womb. It shows up there, because there is a tendency in the soul life to deposit around that organ. The soul life has an issue in that organ—a thought pattern connected to it: let's cover it over, make a deposit—or it gets preoccupied, or we could even say obsessed with producing something from a womb.

Our life today is dominated by sclerosis, breakdown, and depositing what we could call facts, data, stuff. If our whole consciousness is embedded in that, no wonder that sclerosis is the basis for scourges rather than inflammation. Inflammation today is a secondary response to the primary sclerotic conditions of life. In the ancient world, sclerosis was very rare, but decay and inflammation were rampant.

Inflammation and sclerosis are the two poles, and they are the two sequences that therapeutic eurythmists use. A doctor would make a diagnosis and recommend working with these two things, because this is the underlying root of the problem. The goal is for the "I"-organization to learn how to go to these two polarities without obsessing. It needs to go through the catabolic phase and to follow it with an anabolic phase, and everybody is happy because that anabolic phase will be followed by another catabolic phase, followed by another anabolic phase. If that rhythm is established in the soul by the "I"-organization being able to grab onto and then release the target organ, then the rhythm of that oscillation is the healing force. The rhythm has its roots in feelings, the astral realm. However, it is easier to change my thoughts than my feelings. This becomes the work of meditation on esoteric physiology, and this is what I share with you when we do the body scan. The purpose of that is to hold hands with your "I"-organization and say, this is what it says on the schedule—catabolism and then anabolism, and the "I"-organization says, *Yeah, that's what it says in my job description. Okay, let's go into the marrow—catabolism–anabolism, right? Okay, let's go into the liver—catabolism–anabolism, and then the spleen—catabolism–anabolism, right? Okay, now let's go!* You are just doing a checklist of inventory of how your "I" lives in the journey of your blood, from the center of death in the marrow to the spleen, to Saturn.

That's the exercise. It is to let the "I"-organization get it: catabolism–anabolism. *L, A(h), O, U, M – T, S, R, M, A(h)*; thank you in my thighbone. *L, A(h), O, U, M – T, S, R, M, A(h)* thank you in my spleen. When you do this regularly, eventually imaginations start to come to you of things you can do to change your feelings around the issues that are keeping your "I"-organization obsessing with a particular part of your physiology. That is the idea. You consciously use the archetypes of physiology to go into the organs and say this is how it is done. The "I"-organization will greatly value the fact that your personality finally recognizes the extent of its task, which is to maintain balance in your organism throughout your whole life. "I"-organization. It is not the whole "I"; as already stated, the whole "I" has a destiny issue with which it is also in contact, as well as with your Angel, your karma, and the Guardian. That is all also part of your "I," but there is a function of your "I" designed to allow your spirit to enter your body, basically a place where it is alien, because your body is packed with substance. This is not part of the original contract. That substance gives you a particular tweak in your consciousness. It creates pathological attention or, as Rudolf Steiner puts it, a sense organ in the wrong place. That is what he calls a tumor—a deposit is a sense organ in the wrong place. A benign tumor is just a warning, a wake-up call to look at it emotionally. It is also congenital things and the karma of an issue that you need to deal with, which you set up before you incarnated.

The heart descends through the embryo, because that descent of the heart is a parallel of Christ's descent into the Earth—an incarnation of the Sun into the Earth. These two exercises are given simply as a means to touch into an organ, and the sequence with which we go through the body is the sequence of the blood, because that is the organ that the "I" occupies. It took many years for me to figure out this exercise so that it is as free of errors as possible when giving it to human beings, which is always a gamble. But to have the imaginations from Steiner that he gives in physiology to go through an exercise like this, and then finally someone gave me a book—I have forgotten who the man is; today he has a website and does all this psychology that is called a body scan. He gave me the book, and I read it. I thought, *Oh My Gosh, it's a body scan, except he does it like a mechanical thing. You start in your toenail and you just go through and right on up, as if it was a ring going through your body, and then you go out through your head. And it's health-giving.* I knew that there is a tradition in Eastern medicine called, "the inner smile."

You just go through your organs according to Chinese medicine and touch them and say, *Thank you, I love you,* and then you move on. So when I found these ideas I thought, *Hey, this is just sort of an anthroposophic body scan.* But the anthroposophic body scan has these components to it out of Rudolf Steiner's picture of physiology that, I feel, can allow you to make a bridge to the soul aspects of why the organ is not doing well, because we are giving the "I"-organization pictures of how it should work. That is the idea.

Peanut allergy is not the peanut itself, but the mold in the peanut. Peanuts have a very strange, astral way of growing. The flower forms above, and then a little shoot forms that burrows into the ground with the fruit from the flower. The fruit grows underground, but the flower is fertilized above the ground. That is the life gesture of a peanut; the fruit grows underground, and as a result peanuts tend to have mold in them, and the mold does things with the protein in the peanut, which leads to a protein allergic response in the immune system. Maybe when you were three years old, you had to eat a lot of peanut butter, and it made you gag a couple of times, but you forgot all about that. When you are eighteen and you go to a Thai restaurant, you have the peanut sauce, and then you don't feel very well and get sick that night. Seven years later you can't even look at a can of anything that was made in a factory where they process peanuts. We could call that a pathological fugue, a theme and variation becoming increasingly evolved, intense, and active. Then we wonder, how did that happen? I really used to like peanuts.

When I went through college, I worked as a dishwasher in a dairy bar, and every Saturday would be party time. There was a big porch on the back, and all the kids would come and the moms would order banana splits, hot fudge sundaes, and chocolate butterscotch extravaganzas. The kids would be crazy, and they would have big dishes full of ice cream sundaes, which they would never finish. My job was to stick my hand in the dish, scoop out the half eaten ice cream and hot fudge sauce, and put it into a big garbage can. Then I ran the dishes in the dishwasher through the recirculating hot water. By the end of the day, that water in the dishwasher looked like it was made out of ice cream and hot fudge. After work I would wash up, go to bed, wake up in the morning, and Sunday would be the same thing. I could smell the rancid milk on my skin, because I had gotten covered with butterscotch and milk.

When I was going through college, I didn't have any money, and I ate at the dairy bar. I was pantry manager as well as dishwasher. My job was to go into the big refrigerator in the morning and take inventory, and right by the door of the refrigerator was a fifty-gallon dairy can of the best cottage cheese in the world. My habit for breakfast was to get a piece of apple pie, start my inventory, open the lid to the cottage cheese, and put a big pile of cottage cheese curds on the apple pie. That was breakfast. Later, when I found out I had milk allergies, guess what—I had cottage cheese withdrawal. A doctor said, you have terrible milk allergies; you cannot eat milk products anymore. We had three kids, and they basically lived on cottage cheese. I would open the refrigerator and see the cottage cheese, and I was like the man with the golden arm—I need my cottage cheese. I would kill for cottage cheese. That was me for about eight months, until that craving worked its way out of my system.

Now, I can tolerate dairy, but I don't like it. My soul got to a point where my feelings about that dishwasher just got into my astral body. I hated dairy products! And some other person, who was out there in the party room loved dairy products. I did, too, but to me, this was a teaching. Now, I am dealing with wheat issues, because as I grew up, snack time was milk-sop—bread and milk, bread and milk, bread and milk, with potatoes and kielbasa for the weekend. Now that I have worked through milk, bread has become an issue. I can eat it, but I immediately feel the way the mucus is moving, because my astral body starts depositing. I wake up in the morning, and I can feel dull from eating bread. I can eat bread for three days in a row, and then I have to stop. But if I didn't know this and just keep eating bread and milk, I would eventually get very sick and then have an organ removed, all because of being unaware that what was going into my body was creating the problem. That is what I learned in my skinny little world about food, appetite, desire for particular foods, and the physiological response through inflammation and sclerosis. There is a direct connection.

Substance is the corpse of an image. Form is a movement come to rest. What we call a substance is, from an esoteric point of view, actually an activity of beings in the ether world. Thus, it depends on my constitution whether my allergy is a response to the image or to the substance. After a while, however, an allergic response means the substance *is* the image. Some people today are so sensitive to this because they have evolved to this level of sensitivity to substances. It's an evolutionary phase. We actually are supposed to do this, so that we realize there is a whole synthetic world that does not really represent the way things need to go. We will address this further along when we go into miasm, because this is the root of what Samuel Hahnemann, the founder of homeopathy, would call *miasm*. Miasm is a certain constitutional tendency toward a particular illness pattern. Miasm is inherited.

Now let's go through our body scan: L, A(h), O, U, M – T, S, R, M, A(h) in the thighbone. Imagine for a while a fountain of blood coming out of your bones—life and light. (Practice this first.). Next, we go into the capillaries for a while rather than into the spleen, because we will practice Omega now that we have enough organs in between to go through the whole cycle. We move that blood to the capillaries on the periphery of the body. L, A(h), O, U, M – T, S, R, M, A(h). Then, from the capillaries go for a period of time to the digestive area, your intestines. Again, do the sequence L, A(h), O, U, M – T, S, R, M, A(h). From the intestines go to the liver, just under your rib cage on the right. From the liver move upward toward the center of your body, just to the right side of your heart and practice the sequence. From the right side of your heart go upward and spread through the lungs. From the lungs move down to the left side of the heart and go through the sequence. Now move from the lungs up into the aorta, then down to the body. Again, do the sequence L, A(h), O, U, M – T, S, R, M, A(h). This is the lower cycle. After finishing this, rest for a while in silence.

Now a quotation by Rudolf Steiner from *The Reappearance of Christ in the Etheric*, lecture 2:

> You must learn to comprehend the spiritual worlds in your "I"-being through your "I." The "I" has descended and is near at hand. It must no longer be sought in a world of rapture outside of consciousness! Herein lies the greatness of the event that took place at the time of Christ: that through the natural evolution of human faculties, the

old relationship with the spiritual worlds was lost and the attainment of "I"-consciousness was achieved, but that it was also possible as a result of this to gain consciousness of these spiritual worlds within the physical world.

That is the big key. Within the physical I can experience transcendent consciousness. I do not have to get out beyond my capacity for understanding. This reference is a Rosicrucian gesture.

> Christ thus became the mediator of the spiritual worlds for those human beings who have reached such a stage of development that they can, in the "I" that lives on the physical plane, gain the connection with the spiritual world.

This means the penetration of the physical to understand the spirit. This is the Rosicrucian stream, we could say.

> Only after the event of Golgotha did Christ become visible in the atmosphere of the Earth, because it is since that time that he has been present there. One who was experienced in clairvoyance during pre-Christian times knew that the time would come when this would occur. Change the disposition of your souls; do not believe any longer that human beings can ascend normally to the spiritual world by being enraptured; rather believe that through the development of capacities inherent to the "I," and with the help of Christ, you can find the path leading into the spiritual worlds.

The capacities of the "I," are the "I"-organization that monitors activities in the physical body.

That is the theme with which we are working. This capacity has descended into the physical body; in the past, it was still in a spiritual condition. That is reflected in the formation of the embryo. There are two polar states in the embryo. Figure 12 (next page) shows a cross-section of an eight-day-old embryo. In the center we see what looks like three worms or caterpillars. Those are known as germ layers in the embryo. The upper one is the endoderm, a layer one cell thick connected to the yolk sac. The yolk sac is nutrition provided by the mother organism. As soon as that zygote implants in the uterus, these layers form—two

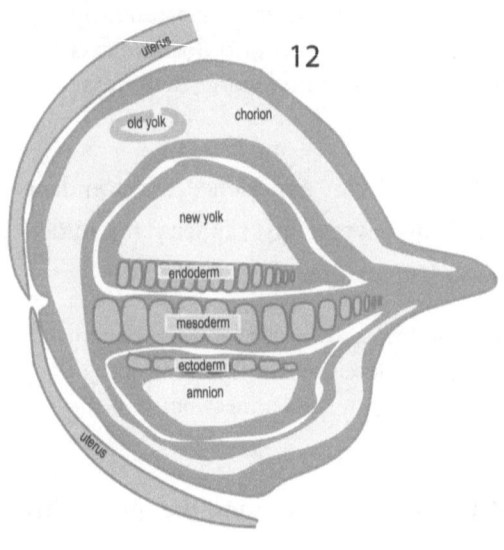

to begin with. One is the yolk sac, and the other is the amnion. The floor of the yolk sac becomes endoderm. All of your metabolic system, the digestive organs, will develop from that layer of the endoderm. The yolk sac is nutrition for the young embryo, so the basis of your digestive organs comes from the nutrition in the yolk sac for the young embryo. That is the one, what is known as a sheath; that is the yolk.

The other down below, is amnion, the large sheath filled with fluid that surrounds the whole embryo on the periphery. The floor of the amnion, however, is a little layer of cells that forms what is known as ectoderm. All of the organs of your nervous system, which include most of the sheaths of your sense organs, develop from the ectoderm. Rudolf Steiner calls that the nerve–sense pole, because in the embryo all of your nerves and senses are developed from that ectodermal layer on the seventh day, when you couldn't even see this embryo. These two layers, ectoderm and endoderm, are an inheritance from the animal world. Some animals exist only with ectoderm and endoderm. They have what is called *mesenchyme,* a sponge as a filling between the two layers, made of little bone-like formations embedded in jelly.

Through evolution this jelly evolves into the mesoderm, but in lower animals it still serves as a kind of filling. In the higher animals the mesoderm is remarkably creative in forming muscles, bones, connective tissue, and organs. Yet, in organ formation, there is a tendency for the sheath on the periphery of an organ to come from an ectodermal layer, and the vascular structure inside an organ to come from the endodermal layer, and all the forms in the organ arise in the mesoderm. There is a tendency for this to happen, not exclusively, but it is a pattern.

In the embryonic higher organism there are two germ layers, the ectoderm and endoderm. Where they touch they secrete into each other, which stimulates cells to migrate into or out of that area. From being an established cell embedded in an organ, they de-differentiate; they change and become amoeboid, and with that mobility they migrate between the two layers under the influence of the secretions. In this process they are involved in the extremely active layer called mesoderm. That is what you see in figure 12. There is the yolk sac with the endoderm, the amnion with the ectoderm, and the mesoderm forming in between.

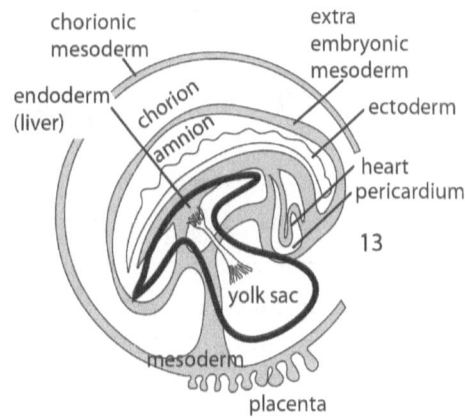

In figure 13, we see amnion on the outside forming a sheath, and then it says endoderm (liver). And you see that the dark line there is the yolk sac. This is maybe three days beyond figure 12, which is perhaps day seven, and figure 13 is perhaps day ten. What happens is when the mesoderm evolves, it just explodes with organ-forming potential. The backbone starts to form and all the nerves along the backbone start to form, but they form out of the somites, out of the mesodermal activity that grabs hold of the ectoderm and says, do this. And so all the organs form under the influence of the mesoderm. The mesoderm acts as a new, extremely active etheric layer that can actually change the organization of cells just by contact through secretion. It is a miraculous thing.

What you see in figure 13 is the yolk sac, and on the left is endoderm (liver). You see there are veins from the yolk sac going toward the liver. Those are known as vitelline, or yolk, veins. What starts to happen at this point (days 10, 11, 12) is that the mesoderm becomes so active it starts to stimulate secretions that cause the ectoderm and endoderm to explode

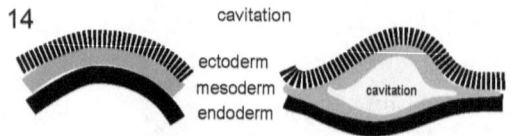

14

and proliferate. When that happens, there is a lot of what is called invagination or cavitation.

In figure 14, we see an example on the left of the three layers—ectoderm, mesoderm, and endoderm—and on the right are those same three layers under the influence of cavitation. In the mesoderm is a spontaneous formation of inner surfaces, the process of hollowing out solid blocks of tissue. Steiner would call it *astralization*. In lower animals astralization is created by ectoderm and endoderm layers growing at different rates. This results in folding known as invagination. To a certain degree, that is the case in a human embryo. But in very critical areas, there is not this folding per se; instead there is a spontaneous formation of inwardness as cavities form in solid masses. Steiner would say this happens because the soul is now entering the body.

With the action of the soul, astralization creates hollows and looping patterns. This is a useful picture, and if we want to find the source of that in Rudolf Steiner's work, we have to go out to the planets, because the planets are where the looping happens in space and time. In Steiner's thinking, that activity of planetary looping is the source of the forms that later become organs. He calls these patterns etheric formative forces. There are spaces created in time and space by these movements of the planets. From Steiner's point of view, that is the source of the patterning of forms that later become organs in an organism. He calls it astralization.

The cosmic astral of planetary motions becomes personal astral through the formation of invagination and cavitation through the process of hollowing out. That hollowing out then creates what we would call an organ. Once an organ is hollow, it has to establish a relationship between what is inside the hollow and what is outside. The relationship between what is inside and what is outside is the basis for catabolism and anabolism, for secretions and non-secretions, and for osmosis and diffusion and the various densities of fluids inside and outside and organism. The archetype is that I have a mass that is very active mesodermically, and then suddenly a hollow opens up within it, and from that hollow a pore forms. Things then move in and out of that pore, having to do with the fact that the hollowing out creates a kind of separation from the environment, it creates an organ. That is a wake-up call—an organ-forming principle. In very primitive animals and plants, it is just a vacuole or something that happens. But then as you move up in evolutionary complexity, the body is basically many bags within bags within bags. If you talk to biologists, the crude way they put it is, an organ is a bag within a bag—an inside and an outside with something in between where there is some kind of activity. Balancing what is in the bag and what is outside the bag is life.

You can find cavitation starting once the blastula forms in a human being. The formation of a blastula is the beginning of a kind of cavitation; a gastrula forms, and that is another cavitation. After that, everything becomes hollowing and filling or not; solid tissue opens up and becomes more complex, and then suddenly you have organ systems. This principle is active very early. You could say that the egg cell, under the influence of the moon and the quality of light that moves through the retina of the mother, starts the process when the polar body is ejected and the egg begins to become more organized. Steiner would say that the spiritual human being is there before any organism shows up in evolution. The spiritual human being *is*, and from that level of organization the tissues and organs and organ systems fall out of that spiritual human.

The formation of a gastrula from a blastula is called *gastrulation*, whereby the outside becomes the inside. This becomes the motif of the formation of an organ. Whether the organ spontaneously opens to form an inside, or the outside goes to form on the inside, the result is that, when the organ is formed, I have something outside and inside, and I have something between that allows them to communicate. The function of the organ is to balance what is outside and inside. Then we get catabolism and anabolism as the processes that allow the organ to do that—breakdown and build-up with what is outside and inside and balancing. We could say those are the fundamental movements in an organism.

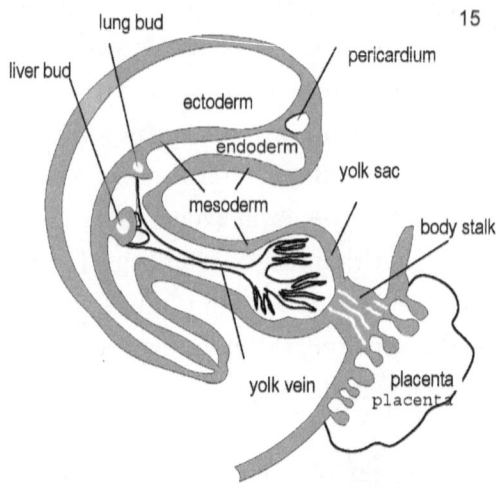

An evolution takes place around the twelfth day. The mesoderm excites everything so that the cavitations start to form. Figure 15 shows the ectoderm as the white area, with a grey area as the mesoderm. Ectoderm on the outside becomes the nerve, endoderm the digestive organ, and mesoderm is the organ-forming, hollowing, a very active layer in between.

At this time, the mesoderm becomes so active that the endoderm (the middle layer) goes into that little bulb on the outside, the yolk sac. We have the endoderm, the future digestive organs, connected to the yolk sac going into the embryo; the ectoderm, the future nerve layer, is on the outside; and in between is a layer of mesoderm. Now the mesoderm becomes so active that it starts to influence the ectoderm to go through a huge explosion of cells. The ectoderm starts to form the backbone and the neural tube, and the brain starts to expand.

You can see this in figure 16. You have the cerebrum, the third ventricle, the midbrain, the cerebellum. That explosion of brain tissue causes the head end of the embryo to bend, called a cervical flexure, as the neck joins the skull. That area of brain tissue grows very rapidly. At the same time, in the periphery of the yolk sac, there is a formation of so-called blood islands.

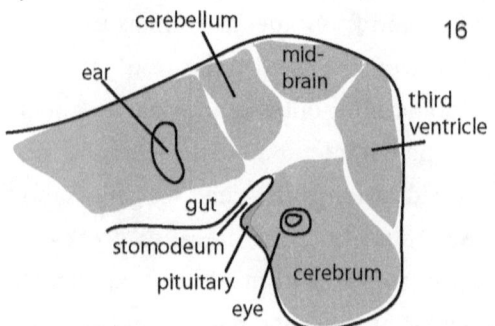

The yolk starts spontaneously to break down, and in the middle of the yolk—in the material of the yolk—little islands of blood form and begin to canalize, or form little canals. As these form, blood platelets either become blood cells moving in or flatten and create vessels spontaneously in the yolk. At that time, they are in the forming blood, and there is a pulse in these little blood islands. There is no heart, but there is a pulse. The yolk starts to dissolve into blood, and the blood organizes into vessels and fluid, and in those vessels with the fluid is a synchronized pulse in the yolk sac. Veins start to form from that synchronized pulse as the veins canalize and go through the stem (figure 15); that's the stem of the embryo. The veins go out of the yolk sac and stream in through the endoderm toward the place where the endoderm and mesoderm are linked.

Where mesoderm and endoderm are linked there is a secretion. Where there is secretion there is cavitation and invagination. In that area, where these two layers touch, folds and openings start to form and will become the future organs of the embryo. Waves of swellings start to happen in the mesoderm and reach out into the endodermal layers, responding to this mesodermal–endodermal touching and the secretions there. In figure 15, we see these veins coming from the yolk sac, going through the stalk and up into an endodermal area that appears as liver bud and lung bud. The mesoderm and endoderm start to interact in such a way that buds start to form and hollow out, and the veins from the yolk sac penetrate into the hollow spaces. And blood starts to pour from the yolk sac into the liver, the future liver. There is a very large plug of material, the great septum in the embryo, that just starts to engorge with the blood that is streaming out of the yolk sac, and as the blood starts to go into this liver area, it starts to excite the mesoderm even more, and all kinds of forms start to sprout out of the mesodermal area, all kinds of folding processes, and they begin to fill with blood. This is the cardinal system, a complete vascular system in the embryo that gets developed and sustains it, and then completely disassembles and disappears by the fourth month—the whole system. It is called the blood sponge.

The blood has to be independent of the mother, otherwise the mother would think that the fetus was a tumor and kill it. It has to be alien protein. There are very delicate exchanges where the capillaries are so thin that the gases and nutrients can be passed through the capillary walls.

At the other end, in the mesoderm—the cranial side rather than the digestive side—the interaction of that mesoderm with the neurology creates a cavitation in the mesoderm on the outside. The mesoderm extends outside, and in that space a little hollow is formed through cavitation of what is known as the extraembryonic mesoderm (rather than the intra-embryonic mesoderm). The mesoderm covers everything; there is outside–inside; wherever there is a little space, it runs from the inside and then goes to the outside and covers the allantois. It covers everything with mesoderm. Simultaneous to the formation of blood islands below, there is a formation in the mesoderm of a hollow space outside the head; that hollow space is the pericardium, where the future heart will be.

In figure 15 you see that little hollow, the pericardium. The brain keeps growing and overrides that pericardium in such a way that the stock of the embryo begins to bend there at a juncture, and at that juncture (see page 53, where we discuss the notochord) the brain explodes and pushes, and right at the spot where there is the organizing center, there is a fold that is created. That is in figure 16. The fold is called the *stomodeum*, the mouth of God, or the hole of God. The third ventricle, the midbrain, and the cerebrum are exploding; the eye is forming, and the stomodeum is forming.

The heart, the pericardium, moves into the area just below the stomodeum. The roof of the stomodeum touches the heart. The heart slides through. The stomodeum is the future mouth of the embryo. The heart basically slides in under your throat and goes down through the mesoderm; as it does so, the ectoderm and the endoderm come together and touch right above that area where that organizing center is in a little pouch called Rathke's pouch, the site of what will become the pituitary gland. The pituitary gland forms as the heart comes into the body and down toward the blood coming up and in from below. At a certain point, the veins come through the liver and the lung, bud up through the mesoderm, meet the pericardium, touch the pericardium, and penetrate and move through it. When that happens, the heart starts to go through an evolution of forming chambers as it descends and rotates down through the mesoderm, pushing organs out of the way. As it does so, it folds in on itself and goes from being a simple tube to being a four-chambered heart. Then the blood comes into it and we have circulation. As the heart turns and folds in on itself, the brain is also active turning and folding (around four-and-a-half weeks) and forming an area known as the limbic structure.

This early neurology is very primitive—just breathing and blinking and stuff like that; it is called the brain stem. However, during the time of the heart's the rotation, there is a simultaneous development of the brain stem; it sends tissue up and forms rotational forms that later become the hippocampus in the area where the pituitary is forming. The formation of the pituitary and limbic structure is analogous to the development of the heart. We have the motif of the heart coming in and rotating and descending, reflected in the neurology by the formation of the limbic system, which is lifting. They happen at the same time. In the embryo, there is this inner relationship between the pituitary and the heart, because the evolution of the four-chambered heart happens at the time when the pituitary and the limbic structure are being formed; they are simultaneous at about four-and-a-half weeks. They are simultaneous developments—one is in the neurology and the other in the meso–endoderm.

This is a key to understanding how we later have to overcome feelings and use our heart and pituitary as a new organ of cognition. This is what we could call the descent of the avatar, the descent of the heart, or the descent of the Christ-being into flesh. It is the descent of the heart. In the place of the organizing center, where the pineal gland will be, there is an organ called the *pronephros*, a bud that later becomes the kidney. As the heart goes in, it pushes the kidney down through the mesoderm. It actually pushes the kidney down, and it evolves into a couple of other organs, until it goes all the way down into the pelvis and then turns around and comes halfway up. Right in that area where the heart comes in and moves things around through the mouth of God, there is a development of the heart as a future organ of perception, the pituitary gland as part of that, and the pineal gland as a kind of ear for the spirit (fourth week).

The kidney starts out in the area of a sense organ up in the head before getting pushed down into the digestive area. Steiner said that the kidney is the main organ for the astral body. The fundamental experience of astrality is having been united with God and now on Earth as a soul, wondering what happened. The kidney was pushed out of heaven and fell. Now you have two of them, but they are little glands, adrenals—Adam and Eve. They get pushed out of Paradise and fall all the way down and get "recalled late and slow."

As the heart moves down, the lung starts out as a liver bud, but then gets pushed up into the nerve–sense pole and starts to build up into the pharynx area developing in the throat; suddenly the lung is a liver that is open to the air. Nevertheless, it forms the same way the liver forms, except that it gets connected to air rather than water because it gets pushed up into the nerve–sense pole.

The lung then has a kind of fear of having to come up, be onstage, and deal with the world, because it is a liver that would rather stay down in the bottom and just float around in the water; suddenly, however, it is open to the outside. Later, the consciousness of that lung becomes *melancholia*. The consciousness connected to the liver, which stays down in the water and just has to cook things, becomes *phlegma*. The *choleric* temperament is based on the heart moving everything out of the way to enter the embryo from a place above the head. The heart says, *Here I come; get out of the way*. The kidney temperament is ruled by *run away!* That's the *sanguine*. Let's go over there and do something else. The sanguine temperament is driven by the kidney—the seat of the astral body. The kidney is always searching for the light it used to have when it lived up by God, up in the head pole as an organ for sensing the cosmos. Now it's down in the back, having been pushed down from the head pole as the heart descended. This sense organ is now settled in the digestive pole. Sanguine people think, *What am I doing here? Maybe I should get a different job. Maybe I ought to move. If I were to move I'd be much better off.*

Moving relieves the pressure of what the sanguine temperament deeply fears: being shamed—being pushed out of one's comfort zone and pinned in the spotlight of shame by disgrace. The phlegmatic temperament is really upset about the day after day after day after day of life. Phlegmatics are resigned to a life of just putting one foot in front of the other. They are afraid of having to spend their whole life down in the basement, working and cooking without recognition. Melancholics fear having to deal with everything outside. They feel that random stuff enters their lungs and upsets them through the fear of being overwhelmed. Basically, they really want to shut down. They don't like all this cat dander, carbon, exhaust fumes, and all the other debris coming into the lungs. Lungs think, *We're part of the body; we're not part of the world out there. We're going to try to make secretions to cover it all.* Because the lungs have to deal with so many different, unexpected matter flowing in with the outside air, those who are melancholic and linked to the consciousness of the lungs develop phobias and tend to be afraid of specific things. When a person has a lung difficulty, it becomes a phobia. They are afraid of cats, because cats have skin that sheds. They are afraid of perfume or perhaps enclosed spaces. The lungs have specific phobic responses to specific allergens.

The descent of the heart is a picture of the formation of organs.

Chapter 8

The Senses and the Life Body

In figure 17, we see the stomodeum, which is a fold. As the brain expands, the area of the mesoderm is so active that these hollows are forming in it, and the big hollow that forms very early outside of the head is known as the pericardium. *Pericardium* means cardio, or heart related; *peri*, around. Our pericardium is the membranous sheath that surrounds the heart. In embryological life before your heart is in the chest, it is outside near the head. The heart has to enter the embryo and move down to find the blood, which has to come in from below in the yolk sac and flows up to meet the heart.

If I were going to design a system, I probably would not design it this way. I would make a bag, fill it with a heart, put some kind of motor on it, and start it pumping. The key to this is the fundamental experience of sensation. The activity of the astral body, or soul, is to receive food, light, sound, and touch from the outside so that those forces enter the organism. We could call this *the astralizing principle*. The vehicle used to transmit those impulses from the outside to the inside is the nerve. If you study the nerves in embryology, there is an area where the nerve is developing; it has to go around, over, and under three other developing organs to reach a target organ. The embryo builds a kind of false target toward which the nerve moves, and when it finally touches that false target, the target disappears and the true target organ arises in the nerve impulses from the sensations on the periphery.

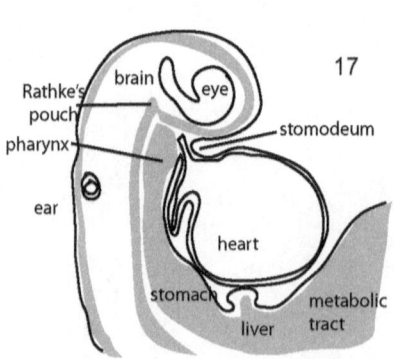

The target organ will later be serviced by that nerve. Embryologists have no clue how this can happen. They have suggested that the chemistry put out by the false target interacts with the head of the nerve. Then they experimented and found that the secretions from the false target that interact with the growing nerve do not tell the whole story.

We generally understand most things in embryology, but when we get down to the nitty-gritty, we fail to understand the real key matters. Rudolf Steiner would say that we don't understand them because we don't link embryology to soul activity, but only attraction and repulsion and balances, ionic discharges, hormonal cycles, and so on. The value that Steiner brings to embryology is viewing catabolic as an astral principle, anabolic as an etheric principle, and the "I"-organization as the guiding force between that balances. The catabolic and the anabolic forces are enacted through the interaction of the nerve in the blood, with the organ formation arising in between as a functional image of the interaction.

It turns out that embryology is fugue-like; the motif of nerve and blood is carried out in every organ. Every organ is an image of a particular connection between the blood and the nerve. The nerve impulse of every organ is the means through which it interacts with sensory activity. That sensation could be, for example, the neurological side of a gland sensing a hormone coursing through the blood or the optic nerve carrying light impulses to the brain. There will be nerves and nerve endings that specialize in receiving—receptors—for that particular hormone, enzyme, or whatever that connects to that, and then there is the occurrence of catabolism and anabolism cycles. The cells respond, begin to metabolize, and then secrete, and those secretions set off a wave of sensation. In this way, digestion and assimilation are fundamentally sense activity. This sensing always happens as the result of an organism, organ, or cell in an environment that is separate from it.

We have glands that secrete substances—glands that have no ducts. They secrete very powerful substances in parts per million. Then we have glands that produce somewhat less powerful secretions through ducts, such as a bile duct into the digestive system. Thus the ones that don't have ducts secrete extremely concentrated substances that are highly organized within the body to create very specific responses, including adrenaline, testosterone, progesterone, melatonin, a thyroid-stimulating hormone, and so on. Those glandular secretions become energetic patterns in the blood. If we took every stimulus response pattern from any gland or secretion—anything that creates a fluid secretion of substances that makes things happen—and could put it all together, watch it flow, and take a picture of it, we would have a snapshot of the ether body. Science would call this packets of information. Material science calls it information. Esoteric science calls it life. It is information, but true information as data has no meaning until the data is cognized as having meaning.

If we analyze the secretions from the thymus gland, for example, we can get a useful picture. The thymus is the gland that trains blood cells to work in the immune system. Basically, the function of a red blood cell is to carry nutrition. There are also white blood cells, which are produced in the marrow, but they are far fewer in number. Red blood cells carry oxygen and dissolved substances in the blood, while the white blood cells are the basis for the immune system and can go through transformations, changing from one form to another depending on their environment. There are special cells in the bone marrow—stem cells—that produce many different forms of cells that can then go through their own morphology when they sense an invader in the blood. Science has a lot of information about the steps in how this happens, but no information about why. *Why* has more to do with soul forces, and *how* has to do only with information and communication. Thus, when there is a "signal," there is a response, but the question is: Where does the signal come from, and how does that organ know to give that signal?

We could say that there is a chemical that hits the organ tells it to give the signal. However, what organ gives that signal? There is a chemical that hits that organ but where does the chemical come from? In chemistry there is a thing called *elective affinities,* having to do with a particular molecular structure that allows another molecule to attach to it. They have affinities that allow them to link up. Where do these affinities come from? Scientists talk about space lattices and fields of dark matter, but still don't have a cosmology for explaining where these things originate. From an esoteric perspective, the order of elective affinities is part of the cosmic consciousness called *music of the spheres,* the consciousness of the Exusai and Dynamis, who created the form principles of planetary movements and so on. In the language of science, elective affinities are linked to the atomic bonding angles related to the space lattice, which has to do with the bonding angles, which in turn have to do with the space lattice. That is as good as it gets, because we have no deeper cosmology for our study of physiology—just information and technology.

In these discussions, I am trying to hold two poles together. One is to present fairly reasonable physiology insofar as I can make that happen as an amateur physiologist. I am trying to link that with the cosmology Rudolf Steiner brought so we understand the aspects that mainstream physiology doesn't understand—to make sense of physiology when we get to deeper levels. This is useful because cosmology has to do with consciousness rather than matter, and activity rather than a corpse. Consciousness acts from the whole to the part, whereas mechanics works with parts connected to other parts to create a cause-and-effect drivetrain. Consciousness is not a drivetrain. Drivetrains are good explanations for mechanics, electricity, and magnetism, while consciousness is holographic. Each part in physiology is a hologram of the whole thing. This is the connection with Bach's fugues; a fugue is holographic. I take the last four notes of the first theme and make a whole fugue out of that, and then every second fugue I add will have that altered eighth note, the dotted eighth note, with a sixteenth note following. That's a holographic motif that is then iterated in multiple ways. At a higher level I will have that as one continuing theme and then insert another theme that developed the motif further in the third one as a new variation on the theme that started in the first theme. It is the same thing repeated in endless variation.

Looking at it from another perspective, there are three levels of fractal dimension. There is the thing that keeps repeating, the thing that is just semi-repeating, and the third level in which the motif never repeats. That is the highest level of fractal dimension and where we get the holographic levels. Technical science touches these fractal levels but has no cosmology of the human being to support the mysteries being revealed; science is cosmology poor. This is because conventional science doesn't relate human evolution to the evolution of the Earth, so scientists have to invent cosmology—big bang, dark matter, string theory, quantum mechanics, and so on in attempts to explain the mystery. Science invents a cause-and-effect cosmology, from which people try to project a model for human existence. Rudolf Steiner said that we do not need to invent cosmologies, because remarkably powerful cosmologies already exist in esoteric realms. To find them, we can look at the evolution of the human embryo.

Now we will direct our embryology toward the immune system. There is a certain stage in the formation of the embryo when the heart moves in and presses in (fig. 17). The stomach and liver are forming along the metabolic tract, and the brain is going up and over. You see Rathke's pouch, where the pituitary will eventually form. Now look at figure 18, showing the next phase and the heart really pushing into the embryo. The organizing area sweeps around, and the heart sends blood vessels into the embryo from its form. It has received blood from the yolk sac and now pulses with blood, and from the heart (coming in through the mesoderm) the arches that will eventually form the aorta start to penetrate in the area of the neck.

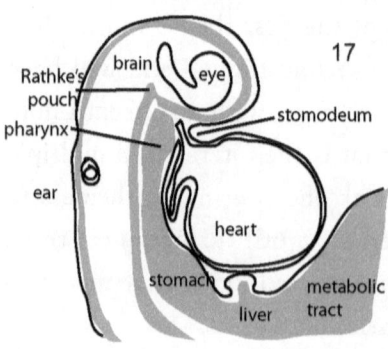

You might picture this as the gill bars of a fish. It's a kind of recapitulation in the embryo of that invertebrate development. In fish, those gill bars do not go beyond that stage and are connected to the circulation of the lungs and heart. They are a circulatory/respiratory organ. However, in the human body they do not function as a lung as they do in fish, but have a different function. The aorta goes in (fig. 18). You see the stomodeum and the aortic arches. They will eventually become the aorta of the heart, the upper part where the blood comes from the left side of the heart, but they are pen-

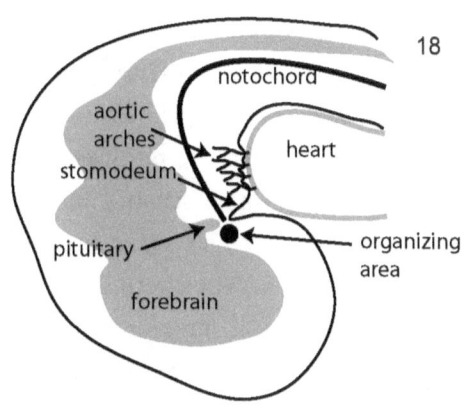

etrating, creating pharyngeal arches around the pharynx. Those pharyngeal arches become the upper palate, the lower palate, the tongue, the hyoid bone, and the larynx—all of the organs of communication for the human. Here we have another connection to the heart, as well as the pituitary, which is in the area of what we could call a future organ of perception and communication involving the larynx, the pituitary, and the heart as the way the world will be created. We could say, it will be sung by us into smoke.

The brain stem is the most primitive element in the brain. There is an area called the midbrain (in the brain stem) that is one of the first areas where primitive neural cortices are laid down. They are little areas in the brain that have to do with the cerebellum. The cerebellum inputs allow me to know that my hand is somewhere in space. This is very unconscious, like the consciousness of a lizard, for example—if it moves, catch and eat it. This is where the pineal gland sits, on top of the optical radiation.

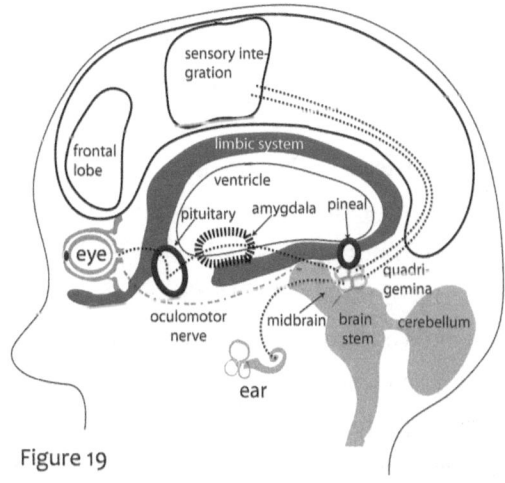

Figure 19

In figure 19 you see the cerebellum, the brain stem, and the midbrain, or reptile brain, which dilates your pupils and keeps your lungs working if you pass out—very primitive but necessary. In figure 19, you see that the eye goes first to the pituitary and then back into an area called the *quadrigemina,* where sensory integration occurs. The ear also goes to the quadrigemina. The pineal sits on top of these. That area is called the optical radiation. Your actual apparatus for seeing is in the front, but the sensory input is integrated in the back in the optical radiation. That area sits below the pineal gland. If you look at the nerves that come out of the midbrain area toward the pineal, they resemble the nerve patterns of the nerves that come from that area forward to the eyes to move the eyes, the ocular motor nerves. But the nerves that would move the eyes and are going toward the pineal are vestigial; they were there at one time in some animals, and they are no longer fully present in the human. They resemble oculomotor nerves and are going toward the pineal as if there were muscles there to move the pineal like an eye, but there are no muscles there to move.

Years ago I read that, and I said to myself that it was significant but I had no idea why. In *Cosmic Memory: The Story of Atlantis, Lemuria, and the Division of the Sexes,* Rudolf Steiner talks about the Lemurians having an eye looking through their skull at the top, which was right where the pineal gland is now. It is right below the fontanelle in the embryo. In that position before birth, the pineal is like a little eye looking up at the warmth patterns and movements of the hierarchies. When the fontanelle closes, there is a fall, and now we look out the front of our skull rather than the top. In case you think this is really just some fantasy, look at the evolution of this in reptiles. Pit vipers, for example, still have this oculomotor innervation to the pit whereby they perceive the warmth patterns of their prey. The warmth-perceiving organ sends signals to the pineal. We lost that when we moved away from snake world.

These pictures are wild. Steiner said that the Lemurians looked at the warmth patterns of the hierarchies through the top of their heads. When you read that you might wonder where their head was. We see that these functional imaginations are written in the lower kingdoms, however, by looking at esoteric physiology, which substantiates so much that Steiner brought by using the embryo and the body as a kind of esoteric script. This is most potent at these very deep inner secrets, which science only glimpses in little, unrelated pieces. They will mention it, saying this is just the way pit vipers have their neurology, but there is little available in contemporary science linking that to a spiritual cosmology involving the human being. Nonetheless, if you want to find another link to the pineal, search Google for "René Descartes pineal." He wrote whole treatises on the pineal gland as a doorway in the brain to the vital force from the Godhead. He described it as a kind of valve that opens and closes based on moral or immoral thoughts. He knew it was at the seat of the fourth ventricle; it is in the fourth ventricle of the brain. Descartes also said that, if a thought is evil, the pineal goes one way and does not allow the cerebrospinal fluid to circulate, and that, if it is a good thought, the pineal goes the other way and fluid refreshes the brain. He has all these theories that the pineal gland was a kind of moral switch in us, letting the vital fluid come up out from this cerebrospinal area into the ventricles and fill them, because the ventricles were known as the womb of the immaculate conception. It was understood at the time that the ventricles and a fluid in them formed the place where the hierarchies create images in us that we call dream and thought. He spent a long time looking at this; it was a great mystery then. What is the pineal gland? And the pituitary—what is that? It is only since 1940 that research on the pituitary has been possible. It was thought to be superfluous for a long time, and now it is the master gland. Since Steiner's time, a lot has happened that he foresaw clairvoyantly; he was way ahead of what people were able to see physically. Steiner said that we need to look at this, and people ask: *Why—because of the Lemurians?* But we could actually say, "Lemurians Я Us."

Figure 20 (next page) shows a breakdown of the pituitary gland, which sits at the base of the skull in a little cradle near the front. Looking into a skull, it's as if there is a little mini-backbone in the bottom of the skull aimed backward, from the back of the skull forward. The pineal gland

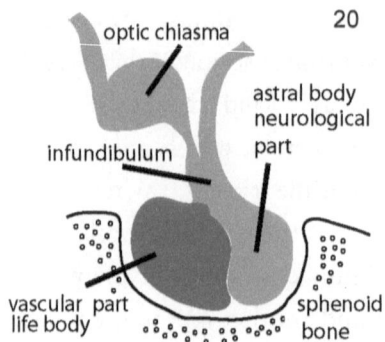

20

would sit, as it were, in the pelvis. The pelvis of this little backbone in the base is the sphenoid bone is the keystone in the arch that holds the face together. The sphenoid creates a little cup that would be equivalent to the pelvic girdle, and the pituitary sits in the little cup.

Look again at figure 19 on page 145. From the eye you see a dotted line going to the pituitary. That is the area we are in here. That line going into the top of the pituitary (in figure 19) is entering (in figure 20) the upper left in the optic chiasma. The optic chiasma area is stimulated by light passing through the retina. It stimulates the optic nerve that goes back toward the brain, but sends an impulse through the optic chiasma downward toward the pituitary. The optic chiasma is where the two nerves from the eyes cross. To see a picture of that, look at figure 21. The two round things on the right are your eyeballs; the optic nerves go back and cross in the optic chiasma; then they continue back to a kind of transmission center in the brain, where they cross again; and then they go all the way back to the optical center on the left. They cross and cross and cross.

In figure 22 (fugue 8), you see Bach's notation for that fugue—the notes going up and descending create the kind of pattern you see in the crossing in the diagram. The form in the music in that mirror fugue reflects the form of the optical radiation in the brain. It is the same kind of wave-like patterning. I wonder if this has anything to do with the health-giving effect of listening to music. The imagination I am trying to

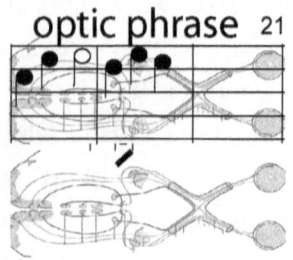

build here is that the geometry he used for the music is the harmony of the spheres that the hierarchies use to create our body. They both use the same source, so to speak, to create form. Later, that becomes the function of the arts in us; when we want to create a new world, we have to explore the creative aspects of form.

In figure 20 (we are back to the pituitary), the astral body, the neurological part, stimulates the vascular part, the life body, with the blood in it. When it does so, there is a secretion of whatever hormone will stimulate the thyroid and other glands that go out through the blood to balance what's going to happen as a result of a sensory impression you just had. When you have a sense impression, nerve impulses go in and all kinds of things happen; you have an automatic governor in your ether body that responds to the sensory impression in kind. It creates a balancing force within you. When we look at the color green, our ether body balances the sensation by creating the impulses to see red. The impact of sensations coming in has to be met by a kind of resistance to keep the organism in balance. If it stays in balance, the organism lives; if it gets out of balance, it no longer lives. I am describing the blueprint for the activity of the glandular patterning of the ether body. It is built to respond to sensory impressions, and this includes thoughts related to sense impressions that stimulate the nerves. The glands receive these forces from the nerves and in response secrete a substance that enables the body to deal with the stimulus. Again, when you have catabolism and secretion, substance falls and consciousness arises.

The source of consciousness related to sensory experience in us comes from secretions as we meet the world, wake up, and things come in. If we consider all those secretions and ask, *Okay, that's a secretion, but what about consciousness arising from the secretion?* Such discursiveness is called thinking and forming an inner picture—producing a mental image that responds to a particular feeling based on a sensation. If the response is strong enough, it becomes a kind of will impulse that stimulates movement in the world. The pituitary gland is the organ that governs these kinds of stimulus–response patterns. In a gland, the neurological–sensory input aspect affects the vascular secretory output side, or life. The vascular aspect secretes in the blood and consciousness arises from the secretion as a deposition. Then I say I am sensitive, living, aware, and thinking. This sequence of responses is the basis for the immune response in the life body through inflammation and sclerosis. In a balanced world, sense impressions come in, and those impressions coming in are not me, whether caused by eating a sandwich or something I see; it is alien to me. My "I"-organization has to maintain sovereignty in the world. If it doesn't, I have a problem.

Rudolf Steiner puts it this way: All of my sense organs belong to the world. They are actually not mine, since it is very difficult for me to be aware in the retina of my eye that light is stimulating the pigment and instigating a nerve response. I can locate my stomach, but if I try to focus my consciousness on the retina of my eye—I really need to work on that. From Steiner's perspective, that area and its response to light patterns of seeing actually belong to the world. Consequently, the eye is like a camera, a device. The ear is like a piano, also a device. A nerve is basically just a transmission line, another device. These devices are on the periphery and give us access to the world. Tasting, seeing, smelling, touching, or whatever allow impulses to enter from the world, and my life body has to deal with it, which it does by resisting the death force in the nerves.

Through the action of the blood, the life body has to resist what enters through the nerve. The first step is catabolism. If my skin gets cut, things outside may get into my organism directly through my blood. It doesn't go through my eye, tongue, or digestion. The purpose of digestion is to break down diverse matter in the world into fundamental conditions so that my blood doesn't have to do that. The purpose of my brain, neurology, autonomic nervous system, and central nervous system is to process input from the world through the senses, memories, and such so that my blood doesn't have to do that. However, things happen that put either sensations or other, more physical contents of the world into my blood without any filtering. For instance, if I get a cut, bacteria can enter, and the perimeter has been violated. When this happens, the blood has to react because the security measures of sensation and digestion have been breached. This is cause for alarm, because the blood is the vehicle of the "I"-organization, and it carries a consciousness of warmth. Anything that enters from outside has an element of cold, because it is not me.

This is what Rudolf Steiner means by "cold." I can be much more enthusiastic about me than I can about you, even if I like you. I could think you are the best thing in the world, but in the end I am number one. If I do not maintain my organism, I cannot love you; that is at another level. Therefore, if there is a security breach, the "I" living in the blood senses that there is cold, meaning it is not me. In the blood the reaction is to bring some warmth. The old term for that among physicians was *calor, rubor, tumor,* and *dolor.* This means that, first, there is heat, followed by redness, swelling, and pain. This describes the progression of infection and inflammation.

The security cells in the blood adopt a kind of amoeboid movement and search for the invader, whatever that is. We can include strange sensations such as smelling rotting flesh or witnessing an accident. Such intrusions cause blood cells to rush into action. If there is a cut, a big influx of life streams toward the cut, and a metabolic process of catabolism starts. It is a mini-digestion that brings warmth to the invader, to immolate, inflame, catabolized, or combust it; this is inflammation. First there is warmth and then reddening. The reddening is caused by all the blood cells carried toward the breach. As this starts to happen, other substances move out through the blood with information, and other cells go into action.

There is a whole hierarchy of blood cells designed to be first responders, second responders, and triage and ER people, followed by intensive care. All of what we have in the medical system is reflected by blood cells designated to do the same thing. First responders are called macrophages, maximum digesters. It is a blood cell that swarms the invader and digests it. According to medical science, the bacilli and viruses and such have a protein coating; that is how living things maintain their identity in their environment. That is how life forms speak to one another in the blood. Who I am is coded into my protein coating. The macrophages are designed to engulf the invader, absorb the protein of the invader, and replicate the invader's protein code on their own surface.

This is the role of the first responders. They go in and grab the invader, ingest it, and then replicate it so that the second responders can identify the invader. Second responders are also modified blood cells that come in to search out things that look like this. Any time the second responders find that particular protein code, they are designed not to replicate it but to get rid of it. The first responders find the problem by identifying the invader. The second responders go attack anything resembling that protein code; this is rampant inflammation, swelling, and pain. Then the third responders come in and mop up. Ideally the inflammation, swelling, and pain recedes and healing progresses from an acute inflammation to a granulation of new tissues in a natural form of sclerosis. Inflammation and catabolism have to go to anabolism that leads to sclerosis.

In the event of *total* catabolism, the responders start eating everything, and you are out of business. It is built into the system—calor, rubor, tumor. A tumor occurs when masses of cells go into an area and bring lymph in to take whatever has been killed and damaged, running it through the lymph, because there are other blood cells in the spleen and in the lymph nodes that take the residue of that thing and neutralize it. Their purpose in the lymph nodes—the blood in the lymph node or the lymph in the lymph node—is to tag everything that has further use and to discard everything else. Tagging allows the kidneys to rid what is no longer useful from the blood. Those tags in the blood let the kidneys know what to do with different substances they receive. We have various communities of these little cells in us that perform these kinds of services that we have out in the world.

The "I"-organization is supposed to take care of this whole process and guide the stages of catabolism that stop and lead to anabolism. After tumor there is dolor, which is pain. By the time I get dolor, the hierarchies

The Senses and the Life Body

have already worked out a lot of stuff. As Steiner puts it, pain is the gift of the hierarchies that helps us attain higher consciousness.

Suppose I have an infection, but I previously had an infection that never worked all the way out to the anabolic phase; so I still have unprocessed catabolic residues in my joints. I did not have enough energy in my life body to reach the anabolic phase from the previous catabolic response. When some of the bacteria that were part of that infection got to the level of catabolism, they formed a little sheath and a little colony out there in the joint. Later, when things improve, they will return and you get another infection from something. This one goes deeper, and you can't deal with it completely, so a little colony forms. And then another infection, and you decide to get some strong antibiotics and whack everything—and we do whack everything, including the good guys.

Over time, a condition of general low-grade infection exists all the time in many people through ameliorating inflammations with salves, ointments, analgesics, and whatever. People are walking around with this soup of compromised anabolic forces in them all the time. It is present in us as little colonies of bowel toxins, blood toxins, liver toxins, lung toxins—basically mucous with nasty ingredients in it. Because of diet and sedentary habits, a certain threshold of toxicity eventually builds up from unresolved inflammatory responses, and the system goes into chronic inflammatory response. This leads to edema, swelling, and things like that—joint pain. In the chronic inflammatory response, the lymph starts storing the toxins in the lymph nodes until there is no more room in them. When this happens, the body is too weak to launch an effective inflammatory response; you cannot get a fever. The system wants to go anabolic, but it can no longer go from inflammatory to anabolic, so it just starts to go anabolic, but without the cognizance of the "I"-organization. Anabolism becomes rampant growth rather than regulated growth, and this becomes deposits.

Thus chronic inflammation—rather than acute inflammation—gradually establishes a situation in which deposits start to form, leading to swollen glands or lymph issues. Now we are dealing with sclerotic diseases instead of inflammatory diseases, but actually the source is a chronic inflammation. In this pattern, the "I"-organization and the life body don't have the power to produce a full inflammatory response but only a low-grade inflammation because there is an overload of toxic residue from unresolved inflammatory responses over time.

We get fermentation rather than true catabolism. When that happens, more toxins are created and we experience the onset of some type of tumor formation, though that does not necessarily mean cancer, which is just one brand of tumor formation. *Tumor* means anything where something is deposited beyond the organism's ability to deal with it. Steiner calls this sclerosis. Often we think sclerosis only as hardening of the arteries because of the term *atherosclerosis*. That is one type, when cholesterol and plaque are produced. In Steiner's language, however, sclerosis occurs anywhere a fluid can no longer hold the substance it is carrying; the substance falls out and becomes a deposit, which is a sclerosis. This includes gout, diabetes, and all kinds of blood issues where something is deposited, depending on who you are, what you eat, exercise, and how kind you are to your liver or kidney. The chronic sclerotic side of this, the chronic depositing side, needs a strong inflammation to bring back some levity force in the body, but if the body is too weak to do it—if there is weakening and succumbing to sclerosis and a hardening, the body has no ability to bring back the levity force.

A change of diet creates a different pH, and toxins start to come out of the tissues and into the lymph. Maybe you take a homeopathic remedy; maybe you start using herbs; you change your diet, and this releases substances from deposits back into the blood. The vital organism, the life organism, wakes up and you start to have an inflammation again, and although you thought you would feel better doing this remedy, you feel worse. In fact, you are getting better. Your life body is getting the toxic stuff out, but when that happens you feel bad because you are experiencing an actual infection as your life body reawakes to the battle against it. If you pay attention, mental images come up that go back to the original infection; your soul will need to revisit this so you can understand the karma of this situation. You may have dreams about it that you can explore. You can go back and research why everybody in your family has diabetes. In my family it is diabetes, so I use that as a case study. Whatever the general vehicle of excarnation du jour, or dysfunction, is in your family, you can usually use these imaginations to go back.

The scanning exercise puts you in contact with certain familial, emotional, and feeling patterns—what Steiner calls *Stimmung,* or mood. Certain moods create an atmosphere in your organism that lead it somewhere in the palette between catabolism and anabolism and back again. Somewhere in there is a key to why in the last round you decided to incarnate into that particular familial situation energetically. It is a rather difficult thing to think about, but when you actually start to do it you can realize that this is the essence of the work that Rudolf Steiner suggests to us, because the implications of not doing it are huge.

> Our contemporary culture is itself creating horrifying monsters that will threaten human beings on Jupiter in the far future. You need only look at the huge machines that human technologies today construct so ingeniously. Human beings are creating demons for themselves that, in the future, will rage against them. Everything that we build today in the way of technological appliances and machines will assume life in the future and oppose humanity in terrible enmity. Everything created for mere utility to satisfy individual or collective egoism will be humankind's enemy in the future. Today, we are far too concerned with gaining useful advantage from what we do. If we really wish to help advance evolution, we should not be concerned with the usefulness of something but with whether it is beautiful and noble... Everything beautiful and noble that we cultivate today leads to strengthening the good on Jupiter; everything that occurs as a result of the egoism and utility leads to strengthening the bad.*

What does that have to do with physiology and esoteric life? The "I"-organization, by participating in this work, has to be given the experience that it is a player and not a victim. We have to grasp the "I" in such a way that we can direct it into the organism to find the pictures created

---

*   Rudolf Steiner, *Guidance in Esoteric Training*, pp. 111–113.

by the hierarchies so we can sift out their pictures from what we inherited from our family. The pictures that the hierarchies created for the organism and the pictures that my family gave me through DNA are not always in sync. Nonetheless, I have to be able to form an organ in me that can tell the difference. The organ I need to form is my heart connected to my pituitary, allowing me to see the quality of movement in the warmth in the world—the nobility and the beauty of things rather than the utility of them. I have to cultivate in myself an organ of the perception of form as a spiritual playbook for the evolution of the human soul. This requires aesthetic perception that leads to moral perception.

Cognitive perception is in the nerves; aesthetic perception is in the feelings; moral perception is in the will. I cannot become conscious of moral perception in my will only through my thinking. This great difficulty leads to significant misunderstandings about the meaning of a human life.* In the cognitive domain, in my nerve, I perceive concepts. In aesthetic perception, my feelings perceive the ability of whatever I am working with to develop further or not. My feelings perceive the potentials in things. Developing such perception is what aesthetics is about. Aesthetic perception allows me to see whether or not something I encounter can develop further. In art I may work on a particular form; something is not right and I get tired and leave it; I come back and look at it and see that a curve is a little off; I have to change it. There is no concept that tells me a curve is wrong; it is a feeling that, if I leave that curve as it is, it does not move within me in the right way. I say to people: When you're drawing, your enemy is a little white space that you can no longer see—a light where a dark should be. You just "grandfather" these random lights into the scheme, and the scheme becomes so precious to you that you cannot change it because you have put the time into it. Darks in the wrong place stand out as an intrusion. Lights in the wrong place sneak under the radar, and I have to bring extra vigilance to them to bring the work forward. So, aesthetic perception in the feelings allows

---

* More on this topic may be found in a lecture titled "The Etherization of the Blood," in *Esoteric Christianity: And the Mission of Christian Rosenkreutz*.

me to feel the balance of the whole field of form. It is a process of training the heart to say, this can develop and this cannot develop, and it is a kind of training in the quality of warmth that is found in wholeness. Trained feelings perceive potential.

In the will I perceive the moral dimension of potential, but I cannot do this with my thinking or even with my feeling, because my thinking cannot penetrate into my will until I learn to *will my thinking*. The easiest way to do that is to train aesthetically, so that my heart learns what impulsive will leads to when it acts alone. When I penetrate my will with my thinking I see the motives for my actions. My heart needs to be strengthened; it needs to live in that perception over a long time so I can work with it, and the best way to work on this is through some artistic practice that trains my heart in the aesthetic domain to pay attention to the moral domain.

Movement comes to rest, and when I turn my soul toward the moral realm, I need to take the form at rest and experience it inwardly, back along its line of emergence, and see it inwardly as movement. That is what we do when we push clay around. We have a form; it is at rest, but I need to change it, so I am going to move the clay around. This is an *inner* movement, similar to the movements I discussed in relation to the Bach pieces. It is an inner sense of the motif's rightness; it is an inner experience of the correctness of the way the flow is happening in music form, color, shape, taste, smell, or touch.

I bring this quotation here at this spot around inflammation, because our culture is too tired to get inflamed. Back in the 1960s, there was at least a little resistance, and Kent State happened, and then Nixon, and so on, and we become so tired of the lying and stupidity that we have just opened the door to cultural sclerosis. Feelings of hopelessness and inability to do anything in the face of the giants— big oil and banks. We feel attacked and helpless. Can you feel the cold? That is what I mean by sclerosis. What can I do?

Well, you can make beautiful things—make noble things. Create rather than consume things. To me, this is what Rudolf Steiner is about—create, create, create, create, create, because what you create will go on and develop into something you could never imagine. Creativity stimulates healthy inflammation that gets rid of the toxins, so that they are metabolized and pumped out. We can be hopeful that this will stimulate healthy cultural renewal.

Chapter 9
Digestion and Emotional Life

I thought we would kick off with a little bit of meditative work, just to create a vessel. Start in the thighbone, do L, A(h), O, U, M – T, S, R, M, A(h). Imagine the flow of blood streaming out from that area and flowing out toward the capillaries to begin with. Put your attention on the surface of your body; the capillaries are on the surface. Practice L, A(h), O, U, M – T, S, R, M, A(h). From the capillaries the venous circulation will flow to the digestive area, so in your intestines do the consonants and vowels (the sequence). Now blood will flow from that digestive area into the liver. In the digestive area the lymph and the blood gather, a sea of blood, and this moves up through the portal vein into the liver. It is on the right side under the rib cage.

Do the vowels and consonants. The liver lifts the blood into a living condition and sends it to the right side of the heart. Do the vowels and consonants. From the right side of the heart, the blood flows up into the lungs and expands. Do vowels and consonants. Now the living blood is oxygenated and flows into the left side of the heart. The quickened and living blood now flows out of the left side of the heart. Do vowels and consonants. The one stream goes to the aorta and descends into the body. The other stream flows out of the left side through the aorta up toward the head, and moves first through the thymus gland. The thymus gland is right below your sternum, where your collarbones meet (pause, do sequence). Blood then flows up to the thyroid right below your larynx (pause, do sequence), and then finally into the brain case (pause, do sequence). Going backward from the brain to the larynx and the thyroid (pause, do sequence). From the thyroid to the thymus below the sternum (pause, do sequence). From the thymus back to the heart (pause, do sequence). From the heart to the lung (pause, do sequence). From the lung to the right side of the heart (pause, do sequence). From the right side of the heart to the digestive area (pause, do sequence). From the digestive area to the center of the bone (pause, do sequence). Express some thanks to the creation for the wonders of this. Rest for a moment in silence. Very good. A good way to start today.

What we are going to look at now is tumor formation. Rudolf Steiner tells us in "The Etherization of the Blood":

> If we cultivate better thoughts we can work indirectly upon the will, but we can do nothing directly to the will that concerns life. This is because in our daily life our will is influenced only in an indirect way, namely, through sleep. When you are asleep you do not think; you do not form mental pictures. The will, however, awakes, permeates our organism from outside, and invigorates it. We feel strengthened in the morning because what has penetrated into our organism has the nature of will... because at night we are asleep regarding our intellect; we become unconscious of what we are undertaking with our will. What we call moral principles and impulses are working indirectly into the will. In fact, human beings need the life of sleep so that the moral impulses we absorb through the life of thought can become effective activity. In ordinary life today, people capable of accomplishing what is right only on the plane of intellect; we are less able to accomplish anything on the moral plane, where we depend on help from the macrocosm.

We cannot change the way we do things just by thinking that we would like to do it. We need some help.

Now I bring this as a picture when we are looking at what a tumor is, because the nature of the will is warmth and enthusiasm, and the nature of a tumor is no warmth and no enthusiasm. So if we want to get some help dealing with sclerosis, we need warmth and enthusiasm, but we cannot do it through thinking. There has to be something else. I will read an excerpt to you from *The Anthroposophical Approach to Medicine*. I will start with a little bit from page 176 and then go to page 179:

Let's consider the sclerotic tendency. It occurs when we consider that the human organism has come from a predominantly water element in embryonic life that we must call the state of the newborn relatively hard.

It has come out of the fluid, even though it is bouncy and juicy and gushy, is hardened relative to where it came from.

Rudolf Steiner saw in this hardening tendency, so necessary for normal development and formation of the organism, essentially the same process as appears after the middle of life as a sclerotic tendency. In youth, this process expresses itself in the normal formation and structuring of the solid organism: the nerve–sense system, the skeletal system, the teeth. The same process that is a healthy one in youth, however, appears as a cause of many diseases in later life. This is because the anabolic process in youth—connected essentially with inflammation—is still dominant. The equilibrium between these two opposed processes, which we call health, is less labile and dependent on age. Either tendency, as such, is pathological, but only when one appears in excess or in the wrong place at the wrong time. This fact is reflected in anatomical data.

The absolute weight of the liver, both in the well nourished and in the malnourished people, is at its maximum in the fourth decade, reaching its minimum in advance age. It is similar with the pancreas. The liver is the central organ of the anabolic metabolism and the main vehicle of the ether body. When we see that in the second half of life, the liver loses about half of its weight without any external cause; we must regard this sort of archetypal phenomena as aging. This fact shows us that, at the beginning of mid-life, the etheric body withdraws increasingly from the metabolism. This, in turn, is because the astral body and ego ["I"] consume and transform the etheric body increasingly throughout life.

This consuming of the ether body by the higher members of the human being offers the possibility of a spiritual development and is the real cause of aging. The metabolic process becomes weaker after the middle of life. If this fact were to serve more as a guide for the dietetics of the elderly, older people would suffer less from the diseases and troubles that largely stem quite simply from the overloading of the metabolic system.

We can ask: What is the metabolic system? Venous circulation is the nutritional, up-building side of the physiology, because in inflammatory and digestive processes, the beginning of the digestive process is an inflammatory experience in the body that begins to break down what is already organized. When I have to digest a sandwich, I break down the sandwich into its carbon, hydrogen, oxygen, and nitrogen components. My body has to catabolize the structure of the food to metabolize it. It cannot metabolize food substances as they are, because if food gets into the blood without being rendered, it would cause species-specific protein responses that would basically kill me. As Rudolf Steiner puts it, the human organism has to resist the world in all these processes. He says that we have to understand that the human being, especially the spiritual human being, has to resist the world. If we do not resist the world, we cannot build ourselves up. An anabolic process cannot develop in a healthy way unless the catabolic process has been completed.

Leaky-gut syndrome is basically a case of incomplete catabolism based on the improper ability of the "I"-organization to deal with calcium. There is a so-called calcium gate in your intestines. It opens between what is inside and what is outside the intestine. Proper digestion requires that this inner–outer interrelationship of calcium, sodium, potassium, and magnesium is balanced. There is a gate in the intestine itself, and if there is an overload of calcium, potassium, magnesium, sodium, or an imbalance among these minerals and metals, the gate opens to allow what is outside to come in and what is inside to go out to establish homeostasis; then the gate closes, and we are back in business.

Mineral nature is the hallmark of lower digestion, but there is a higher digestion that also needs to be healthy. Higher digestion is composed of sensory and emotional experiences. The integration of sensory experience within emotional life requires a higher kind of digestion. Rudolf Steiner calls it *light metabolism*. Science approaches it but would never call it that. When you read the literature, you see that science is playing with that idea, but they could never call it that because they don't have an idea that there needs to be a relationship between the astral and ether bodies; they have no cosmology for it. In the old days, this was the basis in Europe for sending those who had contracted tuberculosis to a spa.

They sent them to the mountains, where the light and the phenols in the air from herbs in the meadow would go into your organism and change the structure of your blood. That kind of sensory–emotional experience is also a digestive tonic—a kind of metabolism, and being a digestive process there is also a catabolic polarity to the digestion of light and emotion. I have to catabolize those as well. If I create an imbalance in calcium through the food that I eat, especially calcium and magnesium—that is, if I eat mainly starches and calcium-rich foods and don't produce the enzymes to catabolize them properly, the calcium gets out of balance with the magnesium in my intestines and the gate stays open. Then there is blow-back from the pressure of what is in the intestine, which includes lymph and all the dead blood cells and gunk that the kidneys can't deal with. This gets blown back into the intestine and creates a toxic ferment where there should be catabolism. It keeps circulating the old fluids. This is called leaky gut.

Although we love our food, the body considers food alien. Hindus have an old saying: Eating is an illness. This is because food taxes the organism and requires the use of a lot of forces to digest the food. Steiner talks about how eating only vegetables requires a tremendous amount of vitality to overcome. As we grow older, we use that vitality for consciousness. When those kinds of diets were developed, the goal was that people would not have anything to do with life or bank accounts or whatever. The idea of incarnation was just to return into consciousness and not have to worry about being on Earth, because "self" is an illusion anyway, so who cares? In the old days, who cared whether my 401(k) application was on time?

When we eat that kind of food, basically we move away from being able to think here on Earth. We are saying to the world, *I don't want to be down here thinking about the world. I want to be up in ether spheres.* With only vegetable food, we use a lot of digestive fire to overcome the food. This energizes the life body when we are young and have good digestive powers; it makes us very healthy. However, the "I"-organization uses those same life forces to form thoughts. This means that we won't have the thought forces available that we need to do the heavy lifting of digestion. This means that, because our life forces are occupied with intense digestion, we will be less inclined to use the "I"-organization for thinking, with the result that we are not as ill.

This is just the way we are designed. I am less ill in my body but my spiritual development goes along the path of the old way, because the contemporary issue for esoteric life is to engage the death processes of thinking based on abstraction and turn them toward spiritual perception. Thus, when I just eat vegetables my consciousness will be absorbed in digestive activity. My thinking will be less discursive and more removed from life on Earth. This is the old consciousness of cultures of the cow. Today, however, most human beings have landed in an abstract world where the life body gets killed by the thinking required of the "I"-organization. This is a natural pattern, especially as we get older, because there is a natural tendency for diminished life forces to be overcome by the "I"-organization. There is a natural tendency for catabolism to dominate as we age, because catabolism brings consciousness at the expense of the life body. If it becomes extreme, we get sclerosis because of the depositing that catabolism does.

If I overload my body with food, it prevents me from breaking down the substances right way. If I am no longer young and robust with a strong digestive fire, and in addition if I have a sedentary and cerebral lifestyle, I use a lot of forces for thinking that will not be available for digestion. People who have aged and live a cerebral, media lifestyle tend to eat foods that provide quick energy, such as refined carbohydrates. Indigenous cultures get around this problem by eating pre-digested substances. If you want to eat a lot of vegetables, ferment them, because the bacteria relieves the digestive system of some of its work. Indigenous cultures based on dairy will ferment it into yogurt or buttermilk. Those that live on beans eat miso or tempeh. Fermenting organisms break down the food for you, so you don't have to use as much digestive fire. If you eat nothing but raw food, you get a lot of help with the enzymes to keep your body healthy, but your own body has to do the work of digestion.

The forces that would be available in later life to develop consciousness have to go to digestion instead.

By taking in substances, the world enters the organism. The substances themselves can be thought of as an earthly nutrition stream. That stream creates secretions in glands and organs as the organism seeks to balance the impact of the substances and maintain sovereignty over the influx from the world. Consciousness of this is an inner experience of resisting the world. This level of consciousness is subconscious and dominant in the autonomic nervous system.

Taking in sensations from the world also creates secretions and stimulates the glands and organs of the body. The forces from these experiences can be called a cosmic nutrition stream. This is also a form of feeling resistance to the world, but the soul has access to this feeling of resistance through the production of inner pictures. This level of consciousness is awake and dominant in the central nervous system.

This is probably one of the hardest things to understand (other than the etheric body) that you will find in Rudolf Steiner's work, because we have a mechanical way of looking at what light is in physical science. However, it is interesting to compare an image from the Danish physicist Niels Bohr (1885–1962), founder of electron shell theory in physics, to Rudolf Steiner's image of ancient Saturn. Bohr said that the electron in an atom is at a steady state orbiting around a nucleus. If the substance in which the atom occurs is heated, the electron begins to absorb the energy from the heat and, at a certain point, will jump from its native shell to the next shell available. The heat creates an oscillation in which, according to Bohr, the atoms move out of the original shell and then collapse back to the original shell in an oscillation. When the electron returns to the original shell it gives off light as a transformation of the heat it absorbed in the expanded state.

Now consider the picture Steiner gives in *An Outline of Esoteric Science,* in which the Thrones radiate warmth (dark, or uncognized, will) toward the Kyriotetes, and the Kyriotetes return that warmth as wisdom (the light of understanding). This is a very similar picture in both physics and Spiritual Science. Considering the electron shell theory of Niels Bohr and the account of ancient Saturn in *An Outline of Esoteric Science*, we see they are very similar images—except that in Steiner's picture beings act and in Bohr's picture there are only abstract forces. That is the issue, because we are now turning toward the whole question of what we could call *cosmic nutrition,* the transformation of the senses and the emotional life. The light of the esoteric student isn't photons; it is understanding. Photons cast shadows; the light of understanding casts no shadows. The light of the physicist is derivative; the light of an esoteric student is primal.

In a mechanical model of digestion, there is no room for the emotional life, but only the electrons that characterize the various qualities of molecular decomposition and synthesis. Nonetheless, there is a vector in there of the emotional life and cognition, and it is not recognized directly by conventional science, but is often recognized as a scientific conundrum arises. A common refrain in good science: We don't know why this is, but it's the reverse of what we thought it would be—especially when dealing with nutrition. This is because there is another stream of nutrition that has nothing to do with substance—or rather, it has to do with substance as an activity, before it manifests, so to speak. We have a stream of substance coming into us that is pure potential, in that it is not manifested. This is hard to understand from a physical–chemical point of view, but as you work with it, as you meditate with it, you start to get pictures of what this could mean. What is this other source of nutrition that is not manifest? Where does the light come from? The light released through molecular discharge is the corpse of this other light.

Have you ever fallen in love? What meter could be used to assess the quality of energy that comes to you or from you to the other? You cannot quantify it, but it is certainly energy. It is enthusiasm and warmth and good will. All you have to do is learn something you didn't understand; a light goes on, and your energy becomes greater. Is there a meter for that? There is none, but we can say it is light, because if I take that "aha" experience of understanding, learn something, and then amplify that understanding, it becomes an *enlightenment.*

Digestion and Emotional Life

According to Rudolf Steiner, a tumor is an intense sensory experience in the wrong place. This would be true anywhere in the case of a "cold disease," any disease that involves a hardening or deposition—gout if it is uric acid from protein degradation or sugar if it is diabetes; it is calcium in the case of arthritis or fat if it is atherosclerosis, the plaque from fats. Something is deposited. Steiner calls fat "solidified cosmos." So, I have to hydrolyze it. I have to render it back to a levity condition. There is a hormone therapy for diabetes using a luteinizing hormone created by the sheath in a pregnant woman that regulates fat uptake in the fetus. It is found in the urine of pregnant women. Europeans use it to affect a receptor in the brain that controls appetite. People are put on 500-calorie-per-day diets with no hunger pangs, and their blood sugar goes way down, because they metabolize their own fat. Once they reach a certain goal, they gradually return to other foods. This is what hormones can do.

In the limbic structure (between the hippocampus and the thalamus and the hypothalamus) there is a sequence having to do with aggression, memory, and a kind of regressive memory accompanied by stress, anxiety, and anger. That gets in the way of light metabolism, because it is hardwired right into the neurology of the brain in the limbic structure, which is the so-called animal brain, or emotional center of the brain. Disturbances there include not only tumor formation but all sclerosis. All these things—arthritis, diabetes, fibromyalgia—are sclerotic. We have all these different names for them, but behind them it's all the same pattern of deposition through loss of levity forces. I say "tumor formation," but not necessarily cancer. A tumor is basically a little embryo without a womb.

Esoterically, the purpose of inwardly working with pictures is to create a "spirit embryo." Through esoteric practice, we are trying to give birth to a new self, and we have to build the organs of that embryo—*chakras*. These are the metabolic organs of the new embryo. If we do not do that, we start making a little embryo someplace where it should not be. Tumors have a very strong tendency to start in the venous circulation, in the anabolic side, on the surface of an organ, and then they go in. That is a pattern similar to the blastula being implanted in a womb. It starts on the surface and goes into a vascular structure looking for nutrition. It then commandeers nutrition in the organ for itself just as an embryo would. That is the pattern. In the venous circulation of an organ, depending on which organ I have chosen to be the womb of this thing, I form a little surrogate embryo in keeping with my emotional patterns, belief structures, and temperament.

I could also add the strategies for risk and reward that I inherited from my family of origin as ways of being in the world. Some people grow up in families in which it is the lung that gets degraded through stress. Some people grow up, and the liver is the organ in which anxiety is stored. This can even be cultural. Involved in this is the karma of what you need and decide to go through. You incarnate in a culture that has a particular diet, which can stress an organ with too much salt or too much fat or alcohol. That country may also have cultural practices that stress a particular organ. You may also grow up in a family that stresses that organ, and thus that organ is at even greater risk.

When your metabolism becomes less active in later life and your etheric body is used by your consciousness, you no longer have the abundant energies of youth available to build up that organ to ward off stress; thus, you start to become sclerotic in that organ. Wherever that is, it becomes your issue. It might be a tumor, but it could be any deposit. Over time, the inability of the "I" to bring light and warmth into the blood causes a general stagnation that results in deposition. This would be edema if the deposition is water, which would most likely be related to the thyroid. The deposition of water through edema is a precondition often used to diagnose a lack of levity in your fluid organism. This may be followed by deposits of uric acid in the water, by fat deposits, or by bone deposits. Alternatively, it could be albumin in the urine, too much protein in the urine, and so on. All of these have behind them the inability of the "I" to bring light and warmth into the blood to create a condition of levity there. This has a great deal to do with the "I"-organization, and becomes a picture of the relationship between how I use my thoughts to connect with my feeling and will. The pattern in my "I"-organization gets its key from the way my "I" is trying to provide my body with pictures, or "imaginations," of my destiny.

The "I"-organization is a function of our "I," which wants to give our body the mandates, pictures, or imaginations of why we came and what we need to do. The "I," which is struggling with personality issues, wishes to do that, because the "I" is in contact with our Angel, reading the book of our whole life's destiny. That is the cosmology. But our "I" needs a vehicle to convey this to the body. This is the "I"-organization, but our soul is between the "I"-organization and the body, using up the forces of our life body through cognition and weird feelings. Our "I"-organization wants communication between the "I" and the other members of the body. The "I"-organization could go into the elemental body and do that, but sometimes it has to work around dysfunctions in the astral body's relationship to the life forces. The astral body and the ether body come together in that realm, and the glue that holds them together is what we call dreaming, which is how they communicate. Dreaming is deeply connected to the feelings we have or repeat as the key to understanding how we metabolize light.

Compare the images on page 170 (figs. 23 and 24). When an organ metabolizes, the substance falls out and the levity rises (as discussed in earlier chapters). When this happens, the levity force is available for thinking, part of the formation process of the substance. That levity force is not random but organized, because it is composed of imaginations of the hierarchies in charge of creating substances and moving those substances in the economy of nature. The organization of the forces of the substance is present in what we could call "intelligent potential."

The substances of carbon, hydrogen, and oxygen are kept in a levity state within plant sap by the action of the plant's life and the capacity for the life to interact in chemical affinities. The life ether and the chemistry, or tone, ether are signatures of the two highest ethers. The substances of the plant are imbedded in the sap solution as the potential patterns by which sugars manifest. In the plant, however, these substances are still

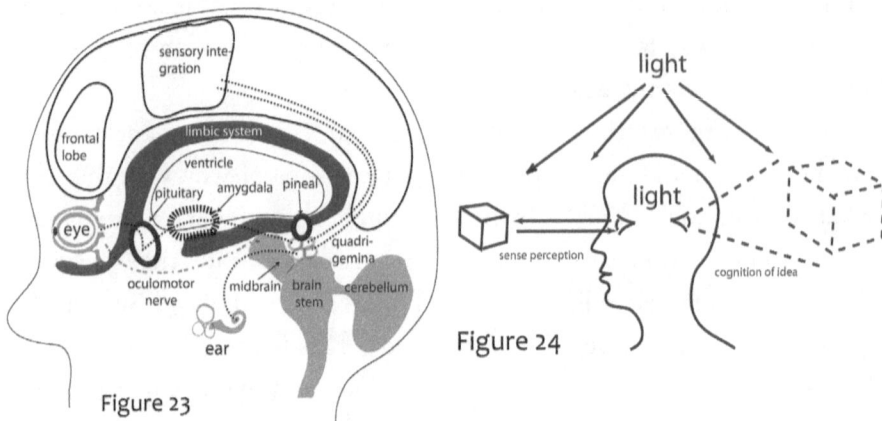

Figure 23

Figure 24

engaged energetically in the life of the plant; they are present as sentient, or sensitive, potentials for attraction and repulsion.

These patterns of attraction and repulsion, their geometry, is the music of the spheres, in which the light of the cosmos is organized into patterns that Rudolf Steiner calls *imaginations;* we might also call them wise ideas, representing the potential for a substance to form without actually manifesting. They represent energetic templates of what could be a substance.

When I look into the world and have a sensory experience, behind what I see are imaginations of very wise, formal movements that eventually manifest in the sugars and saps and carbon and whatever of the plant I'm looking at. If I look at a leaf, I see the corpse of a whole process of becoming that starts with an idea in the mind of God, through movements and activities—we could say energy, or in esoteric language, the light that casts no shadows, the being of the light of intelligent understanding. The light we see and can measure with a meter is a shadow, or corpse, of that light. Why? Because it moves a meter. Photons were used by physicists in Thomas Young's interference, "double-slit" experiment, which created uncertainty (fig. 25). It was decided that photons have some type of interaction with consciousness—but it is not consciousness, which is affected by the primal light, which casts no shadows but is filled with imaginations and creative ideas.

When I have a sensory experience, my soul goes out to that experience (the little cube in fig. 24); I go out to the thing, my soul comes back, and (like an Archai) tells me there is something out there. I have an experience of that Fall. Then, like an Angel and an Archangel, I can interact with that phenomenon and name the thing.

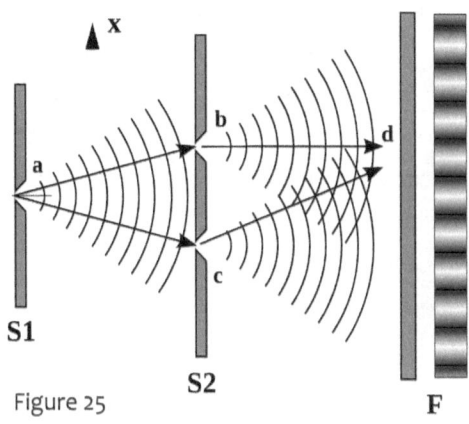

Figure 25

Based on my Archai experience, there is a fall of the phenomenon into the realm of Lucifer as a corpse of the imaginations that gave rise to it. I do not have direct access in my consciousness to the activity of beings of the primal light. In my consciousness, that has all collapsed into substantiality; I take the light that comes off it, and my thinking is based on sensing that corpse. I am free to do that, but in my organism that process involves "killing" the primal light into the thing out there. The Archai experience is in need of redemption by humans who make themselves aware that behind the corpses of the natural world lies a vast realm of spiritual beings acting creatively in collaboration. So, the Archai experience casts the creation down, and as it gets cast down it is moving toward substantiality. I experience the sense world as a substantiality. The predilection for doing that does not allow me to see the action of the hierarchies in the corpse, until I train myself. It is just a habit. We call it seeing and hearing and tasting.

The kind of thinking we do is the key. If I learn to stop thinking thoughts and instead experience myself as a *thinking verb,* I experience inwardly that this thinking is more like will instead of being filled with thoughts that I cleverly arrange as the corpses. My imagination becomes verb-like instead of noun-like. When I practice thinking without thoughts, I begin to experience the quality of will in the force of my thinking, which moves into the realm of infinite potential. When my thinking moves into a realm of infinite potential, I can eventually learn to steep myself in that experience.

When I begin to meditate, a lot of feelings about my life and whatever start to come up very quickly, and often they cannot quite be assimilated, because, as I mature as a person, I have become sclerotic in my soul. I have deposits that have ceased to be flexible and I need to go in with enthusiasm for pulling those things out and looking at them. This is the speech of the Guardian, saying I will go down into the place by the back stairs where you have a lot of fixed ideas stored. Your Guardian suggests cleaning out the garage, so I open a box of fixed ideas that looks like a box of dead rats that need to be dealt with.

In the first experience of initiation science, there is a realization that, when I look out at the world, my soul goes out and comes back and tells me there is something out there that is not me. The activity *behind* the form of that sensory impression is included in that sensory experience. The imagination of the being who created the form is revealed to me through the form of the thing, which is a picture of its becoming. Goethe called it an open secret. It is open wide in front of me, but I cannot perceive its secret life of becoming because of my habitually fixed preconceptions about what it is.

At the same time, when I look at something I am unaware of how my inner soul life and the activity of my oculomotor muscles moving my eyes along the form are triggering my pituitary gland to create analogs in hormonal cascades in my being of that sensory experience. This is simply because I think I am seeing *something out there;* I am still in the Archai realm. I am not yet in Angel world, where I could be collaborating with the Cherubim. But when I start to interact with those pictures and try to control them and see how they are in my sensory experience, I began to experience how these sensations go in and move down as hormonal cascades through my body. I experience this directly because I experience shortness of breath when I look at something, coldness in a certain part of my body, or an inability to digest and tolerate a sensory experience. That starts to come into my consciousness physiologically as a picture, because I begin to touch the mystery of why I am, for example, attracted to this thing. *This is esoteric physiology.* The Guardian helps me understand my relationship to the thing attracting my attention, and when I begin to understand it, I build a spirit embryo through my activity rather than having the embryo build a deposit in my body somewhere as the default.

Rudolf Steiner tells us that I cannot do this on my own but have help from the hierarchies, consisting of the fact that, through my digestive experience, when substances are catabolized the dead light that is lifted from this catabolism is released and goes through the blood, and the blood that is living lifts it to become the source of our ability to form inner pictures. Steiner says that there is actually a stream from the kidneys through the heart that bathes the pineal gland in the back. He calls it "kidney radiation." Not only kidney, but also the spleen and digestive organs do this, but the kidney is in touch with it all as the air body. The kidney is the go-to organ for the astral body owing to its connection with nitrogen, the largest component of our air. The kidney acts as a kind of focalizer of this air—light polarity.

Light is the ether of the air as an element and a shadow of light. I am using alchemical terms because this falling substance releases ether force, the energetic patterns of movement of the imaginations that created the substance. It is released and becomes available to me for consciousness. The kidneys gather all of that from the metabolic regions and send the ether force through the heart and up to bathe the pineal, and from the heart upward to bathe the pituitary. Thus, a metabolism of light imagination is released, which humans use for personal feelings and thoughts about ordinary day-to-day matters. We are free to use that released ether force however we wish, but in terms of esoteric development, I can take those light forces and ether force of levity coming from this breakdown, and I can return them to where they originated.

An image from Rudolf Steiner shows that I must give the world back to the gods. I do this by forming, in the place between the pituitary and the pineal, imaginations of the forms. That is the third ventricle known in the past as the *womb of the immaculate conception*. Light from the kidneys comes up to the pineal, and light from the heart comes up to the pituitary. In his lecture course titled *An Occult Physiology,* Rudolf Steiner pictures those two glands radiating into the space shared between them; this is the source of normal day-consciousness. The space between them is the third ventricle of the brain, along with the two lateral ventricles. If I were to pour wax into the lateral ventricles and the third ventricle as a kind of mold and then dissolve the brain tissue around it, I would have something that resembles a little dove. This is where the forces from the two glands meet.

In Steiner's concept, these two glands create a place in us where we gain access to the original imaginations before world substances fell into corpses. It is in that place where we experience the "breathing process" called waking up (pituitary) and going to sleep (pineal); there is a kind of "womb" between the pituitary and pineal. This esoteric womb can be used to gestate a spirit embryo, which we have to feed from one side, the sensory side, with pictures we have penetrated with understanding and movement that we bring to pictures from the world. The purpose of the arts is to learn how the form of something in the world came into being. When we do eurythmy, paint, or sculpture, we train ourselves to experience the creative movements of the hierarchies behind a formative principle; this is cosmic nutrition. If we join that cosmic nutrition with the consciousness of earthly nutrition, it constitutes a good protocol for one side of the entrance into the great mystery temple—the pituitary side of the Jachin pillar.*

If, on the pineal side, I change my eating habits, digestive rhythms, and so on, it changes the burden my kidney has to deal with. The quality of light that comes from my digestive organs rises and touches the pineal, creating a different "womb" structure. This is Alpha and Omega; if you want to do the work, change your diet, what you look at, and how you think.

---

* *Jachin* and *Boaz* were two copper, brass, or bronze pillars on the porch of Solomon's Temple; their esoteric meanings are discussed by Rudolf Steiner in *The Temple Legend: Freemasonry and Related Occult Movements.*

That space between is a vital place where new forces can be taken and given through the portal of going to sleep and waking up. When I take the day's images in which I recognize the formative principle, it speaks to me, and this could be an experience of awakening resulting from whatever compassion I have extended to them. It might be an insight about my own physiology or a challenge in my physiology. It could be many different things from which a kind of moral imagination arises. The quality of this contact creates forms. If I take that quality as an image into sleep, my pineal gland fires off and creates the great spell for healing; it goes into the "womb of immaculate conception."

When I awake in the morning, light streams through to my retina and goes down into the hypothalamus and into the pituitary. My glandular structure is lit up, and this is the breathing of the other side of the womb into my wakened consciousness. Then I have in space, in my organization, potential for recognizing lawful imagination. I then go about my day and these processes build; I become awake to certain points in time that remind me of the experiences during the night in sleep because somehow, in the mystery temple of sleep and dream, I come into contact with a new kind of substance that has not yet fallen. We call that soul substance *creativity*. It is new; I don't know what it is yet, but I know it seems to have something to do with the practice I am doing.

Creative imaginative processes move like this. I do a workshop and present it to people, and they say they think it is more like this or that. I say yeah, it is more like that. This is spiritual research, the process. The exercise of looking backward over the day's activities* occurs right in this womb. If I take images from the world that I have penetrated artistically, this is more potent than taking analytical matters from the way the consciousness is moving today, which is just an analysis based on ones and zeros; these imaginations are not based on ones and zeros but have the potential to transcend anything we can ever know. A human being who comes into contact with the experience of "more than you can ever

---

* See Rudolf Steiner, *An Outline of Esoteric Science*, pp. 318ff.

know" has two options. We can feel like a bug under the shoe of an angry giant, or we can be incredibly inspired to expand our consciousness toward the exalted beings who are interested in our development. Those are the options.

If I don't begin to participate this way, there will be a yearning through my destiny to have a spirit embryo gestating. Biologically, that means a deposition. But as I get older, the practice really needs to move away from the body, because as I age the anabolic forces of the body become compromised by consciousness and turn more toward the catabolic and depositional. I can overcome the illness-forming potential of this by working consciously with the forces of deposition before they manifest as substance, while they are still in the realm of imagination.

This, in a nutshell, is what I am trying to bring here. It is the function of the arts taken meditatively, in a rhythmic way, into my cycle of sleeping and waking. When that happens, forces start to develop in us that go way beyond what we could hope to expect from simple processes of metamorphosis and transformation, breakdown, catabolism, and anabolism. Because we haven't yet begun to touch our humanity, we are not responsible enough. When we are given real responsibility, we should have a good idea of where we will go and why. However, until more people begin to do this work across the threshold, and until our culture moves away from its Ahrimanic, one–zero, menu-driven consciousness, we will continue to believe that it is the doorway. It is indeed a tool, but don't confuse the wheelbarrow for the gardener. I can use a wheelbarrow to push my harvest up and down the street, but it's just a tool. The real key is shifting one's consciousness away from simple analysis and computational consciousness toward imagination. This is the first step.

The second step is *inspiration,* the response of grace that comes from my efforts to change my will, thinking, and feeling through imagination practice, because there are Angels who wish to inspire us to greater technical expertise. There are technical Angels designated to inspire people directly to greater technical expertise, for the purpose of accelerating the physical power of human beings beyond our moral capacity. This is the realm of inspiration, and it needs to be countered by those who once again take the work of the hierarchies directly through their sense organs to form a spirit embryo in the *womb of immaculate conception*.

Chapter 10

The Neurology of Imagination

The following is a quotation from *Foundations of Esotericism*. It is from 1905, and you can hear in Steiner's words that he is grappling with this issue of the pituitary and pineal. At the time there was very little physiological data available to him to do this, and a lot of his imaginations are based on clairvoyant experience. We are going to look at this area of the "womb of immaculate conception" and certain anatomical structures, looking at whether imaginations I hold in my inner eye give me health. We could ask this: Is the body–mind connection real, or is it just a popular myth?

> The organ of sight is somewhat deeper in evolution than the warmth organ. Through evolution, the organs of hearing, warmth, and sight follow in sequence; the organ of sight is only at the stage of receiving, but the ear already perceives—for instance in the sound of a bell—its innermost being. Warmth must flow from the being itself. The eye has only an image, whereas the ear has the perception of innermost reality. The perception of warmth is the reception of something that radiates outward [warmth, enthusiasm, and recognition]. There is an organ that will also become the active organ of vision. This is present germinally today in the pineal gland (the *epiphysis*), the organ that will give reality to the images that are produced today by the eye.*

Reality is the key word. Recall that Steiner's idea of reality is the concept of imaginations behind the visible world. As he says, the pineal and pituitary glands, as active organs, must develop into the organ of vision (the eye) and the organ of warmth (the heart). They develop what can be called the heart–eye. And my esoteric deed is that I have to bring them

---

*   Rudolf Steiner, *Foundations of Esotericism*, lect. 5 (trans. revised).

together. They will not join biologically but remain separated; I have to bring them together.

> Today, fantasy [the ability to form an inner picture creatively] is the preliminary stage that will lead to the power of creation. At best, we now have imagination, but later on human beings will have magical power [i.e., inspiration].... It develops in proportion to the physical development of the pineal gland. (ibid.)

I am sure that, later, he would have added the pituitary. This womb of the future is between the pituitary and the pineal, where the future human heart will have an eye capable of listening. What will it be listening to? The movements of warmth in its environment, which is how we will meet one another. We actually meet each other that way today anyway, but it is caught up with how we look or don't look, or what we say or don't say, and so on. As we become more spiritual, this organ will be a kind of composite sensory organ. The heart will be the organ of perception, but it will use these glands as eye and ear, so to speak.

The creative imaginations of the hierarchies behind the visible world have created the organs by which we experience the world. They are forms of this creation, but we see them as finished. If we dissect a corpse and remove the pituitary gland, slice it up, put it under a microscope, take photos, analyze it, and so on, we will have an image of it, but the image is merely a picture of the activity; it is the corpse of an activity. My physical body of substances is a corpse I can see because my sense organs are also corpses, through which I see the world as a corpse. The physical body, however, is animated by life, soul, and "I" forces that I cannot see, but I can experience them. The danger is identifying the physical body with those forces. The danger is also that the physical body has analog forces similar to the life and soul forces. We call these analogs of life forces magnetism. We call the forces in the physical body that are analogs of soul activity, or cognition, electricity. Magnetism is the corpse of the life body; electricity is the corpse of the astral body, or the corpse of light. It is a shadow of light that is even below air.

The human phantom, by contrast, is also a set of forces, but it constitutes the original pattern of how the sense body was to relate to life, the action of the hierarchies. The phantom was originally meant to be the human faculty to receive imaginations from the cosmos to inspire joy and love in human beings—to sing back the music of the spheres. The phantom body, if totally healthy and integrated, doesn't worry but is life-giving, refreshing, and a source of limitless creative imaginations. From limitless imaginations, through the hierarchies, come forms that slowly fall all the way down and into substances. Then, even below substances, we have the Ahrimanic deception wanting us to believe that the forces animating the body are actually electricity. In fact, they are a form of electricity known as *primary electricity*, because primary electricity is static electricity. Static electricity is composed of charges that attract and repel each other; primary electricity is the shadow of life forces. Primary electricity is a kind of elemental force in the natural world. There is voltage, or power, in primary electricity but very little amperage, or flow. Primary, or static, electricity is a shadow of the forces of life.

Secondary electricity is a shadow of that shadow and the type of electricity used in our electrical devices. In secondary electricity, the amperage, or flow continuously destroys the power stored in its source (voltage). Technically, in secondary electricity production I take a field of primary electricity, the Earth body and its magnetic forces, and I cut through the force lines with spinning, wound wires to create a generator. When I cut through the field lines of the Earth's life body with the spinning wires, I can pull soul forces (electrical fields) from the Earth's life body (magnetic fields) and run the soul and life forces of the Earth (electromagnetic currents) through devices for my own utility, and a particular consciousness is attached to that. It is not very wise, but it is extremely clever. Steiner calls that process "mining the forces of the earth subnature," which has a whole realm of consciousness associated with it, but it is not the limitless imaginations that are brought through the action of the phantom, although it appears to have earmarks of that type of consciousness.

The electric double is Steiner's term for these forces, which are just ways that Ahriman gets us not to pay attention to the spiritual world. If we know that, we use the forces in a different way. If we don't know that, we think that devices that use these Earth forces are the answer to everything. Electromagnetic force (EMF) is just a good tool, but it is not the answer. As Steiner puts it, electricity was the power of the Godhead, connected to the wisdom (Sophia) of the Godhead. They worked together, but when Sophia was put into prison, power was separated from wisdom. Wisdom was held in prison, and power became available to us through the Fall. Now the being who is prince of the Earth runs the power grid. But Christ is the spirit of the Earth; he has connected Earth's destiny to the eternal cosmos through the reintroduction of the phantom into human evolution. Our work is to try to see the difference. A good place to research these matters is in Ernst Lehrs' *Man or Matter*.

I wanted to discuss imaginations, so I had to go into electricity because many of our imaginations today arise from electricity. It is actually difficult to get to an image of the imagination that comes from electricity—cosmic imagination through images that come to us electronically. At best, we get a kind of flickering picture, or image of an image. This is a critical issue, because the real instrument of human and Earth's evolution is the soul, while the body packed with matter uses attraction and repulsion as its operating principle. This is primary electricity; as such, it is a shadow of cosmic life but gets the job done in the physical realm. Nonetheless, even static electricity is merely a shadow; when we take the poles of electricity (positive and negative) into the soul, we get such things as antipathy and sympathy. This is the source of those polarities, rather than the other way around. If I want to develop myself, I have to become aware of the quality of the imagination driving my soul; I have to work far into the future for everyone. Today we have a parallel world being built of machinery and EMF forces designed to give us imaginations that do not lead back to the hierarchical realms. No matter how much we believe it will, it is like the *Matrix* movies; someone always has to put on the cross. In the movie, the one crucified is the new one, Neo, and the new one is the embryo we are supposed to be building in the image of the crucified Christ-being. The spirit embryo has a tough row to hoe.

This is the Dragon ready to eat the child who comes from the Mother in the Apocalypse (chap. 12). We are in the Apocalypse, according to Rudolf Steiner. It is not about to happen but has already happened. The imaginations we see arising from the Apocalypse form post-apocalyptic consciousness. In the big picture, Rudolf Steiner returned with Anthroposophy to show us that here is another way to do this, and he chose the Rosicrucian path of imagination to renew medicine and bring biodynamic agriculture into the world.

Fundamentally, this work involves sensory experience, and when that experience enters the sense organ and falls, as it must, by creating a secretion, the consciousness that arises from the falling of those secretions is not generally available. It simply unfolds the life function of the body. However, if I work on myself I begin to connect with the fact that there are certain patterns in those forces that become observable. If I study and work with those patterns in the body, I am led to the glandular system and the autonomic central nervous system and to how the glands and digestion interact.

The central nervous system is actually connected to my limbs and movements and my more conscious actions. The glandular system is much more connected to the autonomic nervous system, of which I am mostly unaware. In fact, in evolution the pituitary gland and the central organizing center in the head that we talked about in the embryo actually evolves into the autonomic nervous system in the embryo, and I am totally unconscious in the autonomic nervous system; it is *terra incognita*. However, the fact that I am unconscious means that it is permeated with the hierarchies' creative imaginations. My digestion, autonomic responses, and glandular responses are ruled by the hierarchies. Consequently, they are very much in harmony—at least until I insert my wants, desires, and need; these systems then become less harmonious.

When I am young I don't have to worry about balance so much because of the abundance of anabolic forces in my life body; everything that gets broken down is quickly taken care of, because my life body is regulated by these endocrine secretions monitored in the autonomic nervous system. These secretions regulate all of the life processes, because there are even sensations that come from the inside, which we saw earlier in the kidney. But the pineal–kidney radiation interactions are totally unconscious until I make them conscious. Then, as soon as I start to make these subtle-level interactions conscious, I see the esoteric path of going into my own body and helping my "I"-organization understand why I have certain karmic dilemmas. I begin to get pictures of the destiny issues of the True Self. This is magical work involving changing the diet and finding people to help me understand the destiny issues in dysfunction. If you start to do this, you will draw people to you who will help you understand it. That is also quite magical.

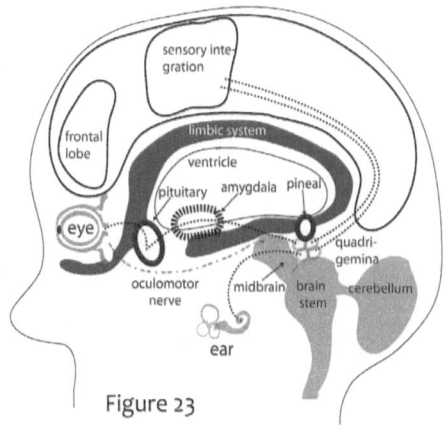

Figure 23

In figure 23, we see the eye on the left and the little dotted line that goes to the pituitary. When we receive sensory impressions through our eyes, they trigger the optic nerve that goes into the hypothalamus, part of the organ called the thalamus. Its purpose is to assess the balance needed with every change that happens in the brain. The amount of neurons in the thalamus is bizarre. We can make a chip with as many nano-circuits as we want and not get near the complexity of the thalamus. Yet somehow everything entering the organ is met exactly by what needs to go out, as the connection needs up and down the line are all continuously worked out. That is the thalamus.

The hypothalamus, by contrast is like a little funnel coming from the bottom of the thalamus into the neurological side of the pituitary (fig. 20). Every time light strikes the back of your retina, the nerve impulse goes through the optic nerve into the hypothalamus and down into the optic

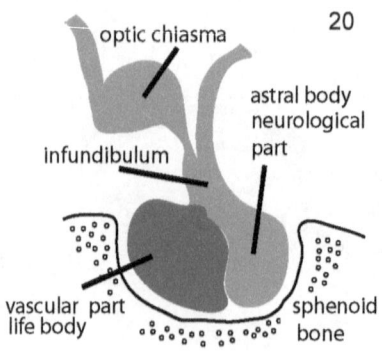

chiasma, which funnels the impulse into the pituitary gland. In these neurological sequences there is a current directed to the neurological side of your pituitary gland.

The neurological side of your pituitary gland is intimately linked to the metabolic (vascular) side filled with blood. When the neurological side of the gland affects the vascular side, a number of secretions are formed. A very important one is thyroid-stimulating hormone. The secreted hormones begin to regulate whatever would be disturbed by the quality of your sensation. This is registered in your endocrine system, because that is your life body, and the purpose of that is to meet immediately everything coming in and balance it.

This is what it's supposed to do. It is a big job, but thank God it is in the hands of the hierarchies; they have a way of doing things very efficiently. Sensory impressions (hearing, touch, taste, feeling warmth, or whatever) are all registered as impulses in a gland and are transformed into secretions that then go through the body to create more and more secretions. This action is called *hormonal cascade*. Every sensory impression stimulates a hormonal cascade of some sort. If we could take all of those movements and put them all in one body we could see, we would see the ether body. We would be seeing the motions of all of the multiple stimulus–response patterns and feedback loops that regulate the life forces.

Thus, the neurological side of the pituitary, under the influence of a sensory impression, causes the blood to release a substance. The nerve kills the blood, a process for which I often use the image of "Cain and Abel." Every sensory impression involves a kind of microcosmic reenactment of Cain killing Abel. Levity forces arise from the fact that there is a secretion; they become available to the "I"-organization for consciousness and to form thoughts. In figure 23, we see an area just above the

The Neurology of Imagination

hypothalamus and the pituitary; it is the third ventricle area that the ancients called the "womb of the immaculate conception." It is bathed continuously by levity forces as the pituitary secretes and releases forces for consciousness; a kind of light show of levity force comes from the pituitary. *Levity* is a code word for forces of motion, which are organized but not yet manifest as substance.

Homeopathic medicine uses substances such as arsenic that would make you ill if you were to ingest such a substance without proper preparation. Arsenic engenders a particular kind of consciousness in human beings. When making a homeopathic remedy we potentize it, lifting the substance toward consciousness and away from manifestation. The idea is that the consciousness and activity of the substance enters the liquid medium, while the substance itself is eliminated. This means that the remedy contains a kind of consciousness of the problem it is intended to remedy. The consciousness of the beings active in creating the substance allows my "I"-organization to understand the pathology in my body. This is discussed in greater detail in chapter 12.

My organism splits sensory input into a secreting phase and consciousness, especially in this aspect of the neurological process. In the brain (fig. 23), the organization of cerebellum, brainstem, and midbrain helps form your base brain, the so-called reptile brain. It governs our autonomic activities and responses such as respiration, pupil dilation, and other primitive neuromuscular responses. The cerebellum integrates proprioceptive bodily muscle movement. This level of brain development is fundamental in lower organisms especially. In the evolution of slightly higher organisms, the brain has developed an additional layer that shows as the heart develops during embryonic life; a concurrent development in the brain is the hippocampus, or limbic structure. It is a kind of layer surrounding the base brain. Look at this evolution in animals; they begin to display social pecking orders, social rituals, joint migrations, mating rituals, and predator–prey relationships that go beyond lizard consciousness. Reptile consciousness dictates, if it moves, eat it. This second layer of the brain developed by higher organisms is the limbic system, which deals with more complex matters, often connected to risk-and-reward feelings and behaviors developed in space and time.

Finally, the cortex or neocortex is the grey matter of the mammalian brain. This is where humans develop language and reasoning functions. It is present to various degrees in more highly developed animals and primates, and in humans it is extremely well developed and articulated. Mammals devote a lot of energy to cortex development. The right side of the human cortex ("right brain") is the site of symbolic, or mystical, thinking. It is the area that is active when we engage in geometrical, creative, and musical thinking and when we form symbolic thoughts. The left side is activated when humans use the language and mathematics faculties devoted to reason and speech.

There is profound development of the prefrontal cortex in human neurology. The prefrontal cortex in humans evolved from the olfactory bulb of lower animals, in which it sends smell impulses into the limbic structure. Dogs have especially numerous neurons devoted to smell, and when their olfactory bulb lights up, they have feeling pictures in the hippocampus, telling them where the good food is and who the enemy is. Smells fill their limbic structure with emotional stimuli. Dogs, as pack animals, devote much of their life to their dog milieu and pecking order, much of it by olfactory means. This is how they survive to adulthood.

The limbic structure is triggered very strongly in mammals by olfactory bulb activity, but in primates and especially humans, the olfactory bulb has been greatly reduced as the prefrontal cortex evolved from that area. The prefrontal area, or frontal lobe, is active when humans are engaged in moral judgment. Rudolf Steiner says that morality is a kind of smell, or fragrance. In the human beings, brain development is linked to deep modifications in sensory input. Much of the brain's neurology is connected with these higher functions, and a lot of energy is given over to the development and functionality of those centers. Intense thinking burns a lot of calories, even though it doesn't involve much muscle activity. There is a tremendous amount of metabolic activity in neurology. In the right side of the brain, consciousness can deal with the qualitative types of reasoning. In the left side, reasoning is very logical and quantitative. Limbic consciousness focuses on how we feel about things. When we have a sensory impression and hear something or are touched by someone in a particular way, the sensory organism relays to us certain feelings that keep us moving inwardly. If those particular sensory impressions are repeated, our neurology, which is very malleable, starts to form those patterns at the expense of other patterns.

Feelings are extremely difficult to bring to cognition. Consequently, changing myself presents a dilemma. I can change my thoughts if I find I am in error, but it is extremely difficult to change my feelings. A repetitive experience over time conditions the limbic system by certain feeling patterns, so when I try to experience other feelings, I can't because a lot of my responses are based unconsciously somewhere in my life body. I don't have conscious access to them in my life body, because that is still in the hands of the hierarchies. This is the dilemma of ensouled beings; we could call it our inheritance from Adam and Eve and the Fall. The dilemma is that the life body has over time been affected by the soul body in such a way that we have memories we cannot recall. These body–mind, psychosomatic memories are the fundamental issue in health.

Now we can add another layer onto this matter—the karma of why I wanted to incarnate in my family, where these feelings were with me day after day, year after year. Now they are part of my neurology and life body, and no one can tell me how to deal with them. I can no longer bring those feelings to consciousness; I can't experience them. It is like trying to discover why you cannot move your shoulder. A body worker might not be able to find the actual dysfunction in the body, but you know it's there because you can't move your arm. When you try to find out what it is, you get close to some form of feeling but cannot go there for some reason. In the cortex I can try to reason it out, but a pathology that is not physiological cannot be reasoned and might be totally illogical. *Only you* have that feeling. When I was about twenty-two years old and just out of the Navy, I returned home and an old buddy came to visit me. We were hanging out and I was complaining. I'd had an argument with my father, because that was just part of the way we always got along. We would have beer and a bump and start arguing. I had a shot and a beer and could see that I had to get out of there, so I called up my buddy and asked him to help me. We went out, and he asked me about what was going on. I said it was my father. He said, "Your father? I love your father. I can go to him with anything. I can tell him anything. He never judges me at all." Were we talking about the same guy?

My limbic structure focuses on my *feelings*—they are *mine*. These feelings are much closer to the soul than the cognitive experience of how a motherboard works in a computer, but how does the motherboard work in my soul? I don't know. What day in my life created this feeling that I can't bring to awareness? What incarnation was that?

We have to go back into the feelings and release the pent-up energy patterns connected to them, so that the toxic substances associated with the chronic stress of unconscious negative feelings fall from where they are stored and we become aware of why we had those feelings in the first place. Such feelings are codified in your tissues after years and years of secretions in that organ, your liver, and so on. The feelings become entrained into the tissues. When we come into contact with our feelings as a source of chronic physiological issues, healing requires an effort to create a space in our soul for chronic fixed beliefs to be loosened and brought up for conscious review in our thinking. In the life body, secretions are codified activities that come from the hierarchies originally given to us to keep the body organized. However, through repeated emotional experiences that become chronic conditions, the organization becomes sclerotic. The emotional organization becomes sclerotic. The soul becomes sclerotic, as it were. It becomes sensitive to particular tones of voice or to the color of a person's skin.

I once lived at a tree nursery in New Jersey, where the young man who lived there was a bit unbalanced. When he really wanted to go on a bender, he would go down to the deli and get a six-pack of Pepsi. Then he would go out into the nursery and chug-a-lug six bottles of Pepsi and be off with the fairies for two or three days—and some of those fairies were not good fairies; we are talking about some serious weirdness. His drug of choice was Pepsi-Cola, and whether it was the sugar or something else, his system responded to it in such a way that it altered his consciousness. When he did, you could tell by the way he was walking.

I had a strange experience recently. My mom told me about this thing called a cybernetic turnpike or something. I typed in the address of the house I grew up in when I was three years old in Philadelphia. On the screen I saw a picture of the front window of the house. You can move the camera around, so I rotated the camera, and there was the street where my grammar school was. I walked through the schoolyard and my grammar school, and then I turned and looked back at the house. It was gone! All of the other row houses were there on the block except the one I grew up in. There was just a chain-link fence and an empty lot between two row houses. It was very profound for me. My house is gone.

Then I put in the address of the nursery in New Jersey, and the camera showed that place. It could not go into the nursery because it was private property, but it did take a shot of the property past the deli, and it shows someone walking across the street. I said, I know who that is—the disturbed young man walking toward the deli on the corner in 2007.

I bring these images, because when I want to change myself in the inner life, I can change the way thought patterns are moving, but I need a different strategy to change my feelings. I have to let feelings arise and reexperience them to cognize them anew. However, I don't want them to arise in such a way that they come running up the back stairs and make me crazy; that is the problem in the first place. Thus, there are various modalities that allow us to assimilate those pictures in little bits. Eventually, as they are released slowly from the life body, consciousness arises, and that helps us understand the issues that led to our chronic feelings. Slow release of these feelings through symbolic imagination begins to lead us toward the karmic reasons for our challenges. It takes research to do this, however. As we research, images of things will come that we can use artistically to help meditate the karma of why we needed that particular dysfunction.

In this way, we can use the arts to plumb our depths, allowing pictures to arise so that we can interact with them as good Angels. When we interact with such images, there is a certain realm we can enter—the door between sleeping and waking. I have worked in this way for a long time, even, as I said, as a study to set up this course. I used my own family as a kind of basis for it with the issue of diabetes, which is a scourge for the men in my family. It is also the scourge of our age. Some of my uncles have changed their diet, but the tendency is still there. They look better and have more energy, but there is still something present. They have not assimilated the soul lesson of diabetes, and I asked myself why that would be. What could be the purpose? Then it kept coming to me: limbic structure, limbic structure, limbic structure. I looked at the limbic structure, but I was looking only at the hippocampus.

In figure 19 we see the pineal gland, with a dark little bar (the hippocampal area) going to the left from the pineal gland. The midbrain is touching the hippocampus. The hippocampal area is a development, in the fifth week, of an outgrowth from the brainstem that suddenly shoots up and curls up like a cinnamon-bun. It then stretches out until it looks a little like the legs of the Sphinx in neurology pictures. The pituitary gland is between the legs of the Sphinx, where the paws would be, and the pineal gland is back where the hippocampus begins. The hippocampus forms the base under the third ventricle; it is like the floor of the womb of immaculate conception. This is a critical area. The function of the hippocampus—the function of the whole limbic structure—is to deal with emotions.

There are short-term emotions, long-term emotions, and very long-term, somatic emotions, or memories. The hippocampus takes short-term sensory motor input and damps it down a little so that the immediate sensory motor impulses are filtered out. Some ambient sensory motor stuff should not be retained, so those risk–reward impulses just come up and are allowed to fade from consciousness. The impulse then becomes available for long-term, biographical memory. The degree of intensity of the sensory experience input and the emotions arising from it determine how long-term the memory will be. Thus, if it is a short burst emotionally, the memory will be short-lived. If it is a big deal or repeated, or if it arises from old karma, that memory becomes deeply embedded in other parts of our neurology and starts to color the experience. When an experience is strong it usually creates a pattern in which

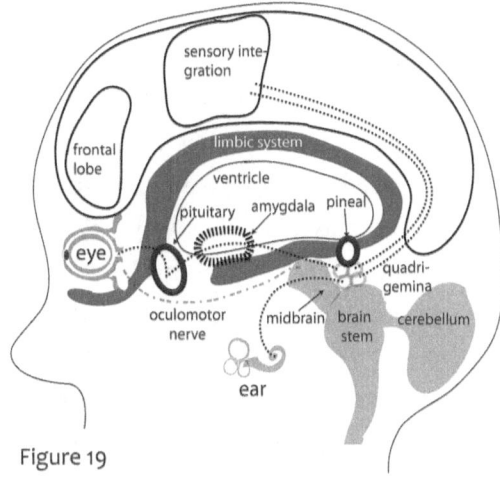

Figure 19

the input is repeated, and the more it is repeated the more it is saved and integrated into our experience of the self. It could be a simple little issue or irritation, but when it is repeated the little irritation becomes a big one that causes big problems. An accumulation of the little irritations are filed away and are experienced by the soul as having strong connections.

In front of the hippocampal area is an area called amygdala (fig. 19). It is on the front end of the hippocampus above the brainstem. The purpose of the amygdala is to act as an overflow for extreme sensory motor impulses flowing through the hippocampus. If you stimulate that area in the brain artificially, aggression and risky behavior are exhibited. The amygdala in the limbic structure is connected to the production of long-term memories based on space–time cues stimulated by the brain's sensory motor circuits.

Connected to the amygdala and the hippocampus is the front of the midbrain, which sits above the brainstem. That area has very primitive neuronal connections to the oculomotor muscles, the muscles that move the eyes. The movements of the eyes are deeply connected to patterns of aggression and risk and reward in the soul. In pack or gang mentality, if you risk looking at me or my girlfriend in the wrong way, you will have a big problem. The oculomotor muscles connected to the primitive neural cortices in the midbrain trigger the amygdala and are connected to deep anxiety states surrounding control issues involving memories of spatial–temporal events in my biography.

Finally, in front of these organs are the pituitary and the hypothalamus, which regulate things such as secretion and the metabolism of blood sugars based on sensory motor responses of risk and reward memories.

Taking this the other way—through the eyes and ears toward the amygdala—I have within me a whole neuronal memory pattern related to very early sensory motor experiences. These early experiences form an organ known as the corpus striatum, which acts as a kind of precursor of the central nervous system. In early learning, that organ registers sensory motor experiences that act strongly in the formation of my nervous system. If something is irritating me during these early preverbal learning episodes—such as a quality of touching or food deprivation—that pattern of anxiety gets stored in the amygdala as long-term aggressive memories. Over time, these memory patterns get buried in the neurology and form a substrate for soul activity later in life. Irritations in the preverbal learning process of repeating without cognition can grow into hyper-alert autoimmune responses later in life. Risk–reward patterns that stimulate aggression can be turned gradually inward into patterns of self-aggression.

I have been trying to understand Steiner's assertion that diabetes is related to an inability of the "I"-organization to enter the body and bring levity. As he describes it, the "I"-organization is weak and it cannot get into the physical body to bring levity to the sugars falling out of the blood. In a normal situation, the "I"-organization brings a kind of consciousness into the blood to deal with sugar, because sugar is a crystal that actually is a part of life, and it can be changed from a mineral crystalline form into a life form by being converted to glycogen, and then it turns into energy. When the "I"-organization is not comfortable, so to speak, in the body, it pulls back and the sugars that go into the blood start to accumulate as minerals. The purpose of insulin in the body is to kindle a kind of "I"-flame to make the sugar come back and be metabolized for energy. The source of insulin is the pancreas.

The pancreas is a key catabolic organ. The inability of the soul and the "I"-organization to penetrate the physical realm—especially in relation to sugar and energy—means that the "I"-organization is pulling back, and the sugars just accumulate because the pancreas cannot deal with them. It would be the same if we were talking about the gallbladder, except that the issue would be fat. The "I"-organization cannot get the gall to render the fat, then fat collects in the blood and atherosclerosis, cholesterol, plaque, and so on become a problem. In this pattern, the "I"-organization pulls back from the body and weakens, and the sugars fall out of the blood and mineralize and there is no way to assimilate them. Eventually, they build a kind of congestion.

What emotional pattern is connected with this? I looked at the pattern of auto-aggression in the neurology. We have long-term memory of aggression and sensory-motor experiences that we cannot control and have been turned inward against the body, which makes it uncomfortable for my "I"-organization to be in my body. As soon as they met you, the men in my family would use words and gestures to put you down. As a kid, I can remember asking myself: Why do we talk to each other this way? I was six years old, and I always got the impression that I was a competitor—a little kid against some 220-pound, six-foot-three Polish guy. I always wondered why they had to put me down. Any of them could squash me like a bug. I was a skinny little six-year-old kid. Yet, they had to put me down with verbal aggression.

Over time I recognized that this created in me a kind of repressed passive-aggressiveness, which just festered in me. But I also had other experiences over time, thank the Lord, and Rudolf Steiner and my dear wife helped me through some of that stuff. I managed to get away from what the female relatives in my family call "the Klocek curse": I meet you; I put you down. Such aggressiveness against my self-image compromised my feelings of autonomy and competency and caused me to dislike my family, which compromised my sense of relatedness. It was a triple-whammy on the ability of my neurology to handle sensory motor responses. I had to assess whether the risk of standing up to the abuse would outweigh the response I'd have to deal with. The usual answer was to stuff my feelings and go away. This strategy eventually compromised my digestive system. Given my particular constitution, those emotional patterns turned into an inflammatory allergic response, because I would always rebel. They would smack me down, I would come back, and they would smack me down again, until I finally left. This was not a solution but it allowed me to assess the pattern more objectively.

I understand this pattern to be the reason I have a strong predisposition to diabetes. I can feel it, but in me it has turned into a chronic inflammatory tendency that balances the sclerotic deposition aspect of the diabetic pattern. In my life, the diabetic pattern has resulted in extreme food allergies, which is a form of diabetes but not "classic" diabetes, because I don't have those symptoms.

I had to go back into my emotional life with certain pictures and ask myself why I get so upset when someone discounts me. I've seen other people respond to being dismissed, and they simply dismiss the dismisser and walk away. I just started cycling, and for the longest time I asked myself why that is. Over the years I have come into contact with it and have been able to find pictures that allow me to understand. What I am sharing with you is that in the realm of imagination you can actually feed yourself pictures that allow you to assimilate the emotional toxins without having to move toward sclerosis.

The turning point for me happened while I was experiencing a deep resistance to Rudolf Steiner's work. When I first found his work, I went whole-hog into studying it. After about sixteen years, however, I hit a wall because of my feelings of resentment toward authoritarians. In a funny way, Steiner assumed the role of my uncles when he spoke so confidently about things. I was struggling to understand and I was in deep inner turmoil and not doing well physically. Then someone gave me a book. The title of the book was *Slavic,* a history of the Slavic peoples immigrating to the US. And it was about the mines; both sides of my family were coal miners. The book said that the people who came from Poland went to work in the mines to avoid contact with the managers.

They went to the worst spot they could find to work and did the best job they could. I recognized that deep within me was the belief that authority figures can't tell me what to do, which has gotten me into a lot of trouble over the years. However, it is Slavic coal miner bravado, something in the folk soul of Poland that resists being a doormat for every invading country since Noah got off the boat. I often say that the Polish national sport is resistance. They are experts at resisting. I see that resistance in myself.

Through resistance in the natural world substance settles out. Sand falls out of the energy flow of a river when that flow encounters the resistance of the bank. And through this kind of settling out, there is the potential for lifting into it consciousness. Every obstacle has a gift hidden in it when we take time to enter the consciousness of the risk-and-reward patterns we place around life's obstacles.

A wise man once told me never to waste a good obstacle in life.

Chapter 11
Emotions and the Will

In the ancient world, it was understood that in certain parts of the body, by focusing on them, one could discern different kinds of consciousness. Through time, numerous people meditated and worked with the human body, and the tradition of consciousness of different parts of the body became known as the system of the chakras. There are various numbers of chakras in the human body, depending on the system, but in general chakras are typically located in areas of the glands.

I have discussed glands, which always have a vascular element—an anabolic, building-up element of the blood and a neurological, astral element of breakdown and catabolism. The catabolic impulse in glands affects the vascular part of a gland; a secretion is made, consciousness rises from the gland, and this is identified as a chakra. The secretion as a substance goes into the blood and creates what is known as a hormonal cascade. People who work with psychology are familiar with the term *emotional cascade*. In the hormonal cascade patterns we see very interesting relationships to those of the emotional cascade.

Over the years I have worked with people sharing aspects of their inner and emotional life, diets, family backgrounds, and so on. I have seen these patterns relate to a system that Rudolf Steiner outlines as what we could call the moral compass in *Foundations of Human Experience*, in *The Philosophy of Freedom (Intuitive Thinking as a Spiritual Path)*, and in *Balance in Teaching*. This is not his term but one I use to describe the seven levels of will. These levels dovetail nicely into the whole issue of hormonal and emotional cascade and the chakras, because *chakra* is a code word for describing sets of patterns related to physiology and consciousness and the catabolic–anabolic relationship.

Figure 27, on the following page, shows a seated person and seven chakras (some systems have more, others less). The lowest chakra here is labeled *instinct,* also called the root chakra. In some systems that chakra is called the collective or the tribal level. We could use the term *kundalini.* The fundamental consciousness of that level is "I win." My cells saying to my blood: Give me your good stuff and take away this waste. There is a catabolic kind of breakdown at a deeply unconscious level in the physiology—consciousness so deep that it is very difficult to gain access to it. In ancient times, it was called the *serpent* that lives in a cave at the bottom and always knows how to win. We call it earth consciousness, the lowest level and instinct.

Rudolf Steiner gives us a picture showing how instinctual patterns come from the actual structures of the body that maintain it against the onslaughts of the world. Wherever that stimulus–response pattern is, "I win you lose" is what he calls instinct at its most fundamental level. However, through use and evolution of the biological phyla, those instinctual patterns at the cellular or tissue level become organs and organized, and they begin to form the basis of the life patterns in organisms that result in certain patterns in one organism and a little different in another.

All organisms tap into the life-body template composed of the totality of the instinctual life that is patterned in complex ways. It is not just stimulus–response or attraction–repulsion. It is stimulus–response that has a stimulus–response balancing it, because a higher organism has a stomach; or maybe, now that it has both a mouth and an anus that are separate, there is a glimmer of freedom from the environmental necessities emerging in the group soul life. In very low organisms there is no mouth or anus but all one thing. Those organisms operate at a very instinctual level. Basically, any one of your body cells is a recapitulation of pretty much any single-cell organism; there is no mouth or anus, and nutrition happens through the cell wall. Inner life at this level is the most basic level instinct.

Once organs start to form rather than organelles, patterns of taking-in and giving-off become codified in structures that have specific

Esoteric Physiology

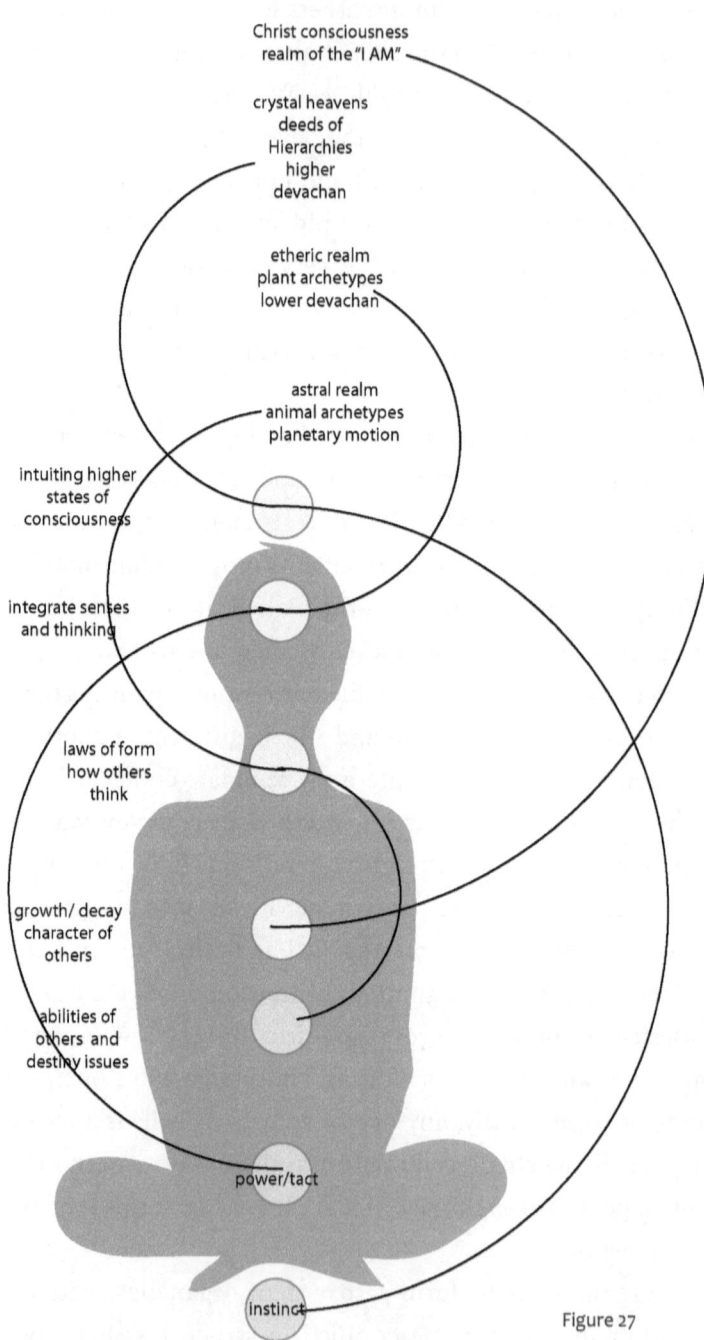

Figure 27

functions within the organism. Those codified patterns (codified instinctual patterns) have a slightly higher level of organization and consciousness. Rudolf Steiner calls the level of will that governs these levels of soul life of organisms *urges,* and one urge is specific to each organism, and another urge is the function of another organism or part of the organism. The simple exchange of gases and nutrition from the cell to the blood becomes codified and organized into patterns of what he calls *urge,* or *drive*. This involves drives toward a particular food or stimulus rather than simply attraction and repulsion of molecular charge fields. Flatworms have a higher consciousness than a single-cell organism, and segmented worms are still higher, because they integrate all of the features of the various kinds of lower worms into one highly functional organ system. When we see flatworms migrating toward a light source, that is urge.

In still higher organisms, urge and drive become more integrated, until behavioral patterns are observed that begin to exhibit what Steiner calls *desire*. However, before organisms exhibit desire, there are many kinds of animals that use urge forces from the natural world to develop a particular niche in the environment, where they become specialists by developing unique adaptive behaviors. Along these lines, I once saw an interesting phenomenon.

My son and I were on a beach and noticed that the crab tracks came out of the surf going to the right, then to the left, to the right, to the left, and so on. The zigzag moved parallel to the tidal line. My son is a photographer, so he took photos of the unusual tracks in the sand. Meanwhile, I was wondering what's going on here. Later, he emailed me and said, "Gee, wasn't that cool? I wonder what was going on with those crabs." There was my question from another person, so I began some research. It turns out that this particular kind of crab has a center for automatic temperature sensing. They feel the very delicate differences in temperature and when they come out of the surf and walk down the beach, the landward side and the seaward side have subtle temperature gradients. The crabs walk along and try to get a little more toward the land where the dead stuff has been deposited at the high tide. But then they encounter a temperature difference that causes them to veer back toward the cooler water. The tracks down the beach show that this inner temperature urge guides their oscillating movements. That sensory motor perception is built into their organism, which responds to natural environmental cues with their activity. We could call this *urge*. Engineers are trying to devise robots that work this way so that they won't have to think but merely respond to stimuli, or what we would call urge. This is the chakra of power, what Steiner calls *tact*.

The chakra just above the base chakra, near the belly button, is governed by power, with the soul quality tact. According to Steiner, the quality of tact in the soul is called *prudence,* which means doing only what is best to keep the organism within the sphere of life. The crab wants to move toward the land, but its organism is warning it not to get too far from the water, so it zigzags down the shore along a prudent line between the safety of the water and the food source on land. Prudence involves assessing risk–reward and the predator–prey relationship of eating versus being eaten. Prudence is codified into intricate webs of behavior among diverse populations of organisms; it is an assessment of power. It is the level of urge and a level of will in the soul. If I start this, do I win?

The first lion on the back of an elephant has a rough time, but the fifteenth lion doesn't have to be prudent because things have changed. The first one has to wait for the prudent moment or it may become substance between the toes of the elephant. This kind of behavior is a little more codified in terms of stimulus–response. Steiner calls it the level of the will in the life body that is urge or drive. In the human this is the belly-button chakra concerned with power and tribal identity as a precursor to more formalized social forms beyond tribal blood-line relationships.

The next chakra up (at the solar plexus) governs our capacity to perceive the abilities of others and how those abilities play into our destiny issues. The qualities of power and prudence in the lower chakra come from activity of the life body. As we move up, the spleen chakra is at the center where the liver, spleen, kidneys, and pancreas work together to digest the world. Those organs have to deal with ensouling, or astralizing, of the stuff that comes in from the outside world. The principle of will in the astral body, according to Steiner, is *desire,* which is an even more personalized urge in the life body in the soul. I can have urges such as hunger is an urge, but hunger for specific things starts to move away from urge as maintenance of the life body and toward wanting specific things that I crave in my soul.

Thirst is also an urge, and has to do with my liver, life body, and glands. But "thirst" for a six-pack of beer or some apple juice is in another realm. Urges become even codified and personalized when an urge starts to be ensouled. This is an activity of the astral body working on the life body in the spleen chakra. In this chakra we have the spleen, pancreas, gallbladder, and liver—the organs that do the heavy lifting of what comes into my mouth, the great orifice that the world uses to get inside us. So that quality of that particular chakra, the way Steiner puts it, has to do with assessing the abilities of others and destiny issues, because he connects desire with *jurisprudence,* a term for having to go to court with someone who is not a member of my tribe.

Prudence in the life body involves heeding risk–reward: Do I beat this guy or does he beat me? Jurisprudence: Can I get away with this or not? Behind jurisprudence is a much more codified set of "laws" that move the soul away from the one-on-one level of existence of the life body and into the vast world of the soul drama of desire. The force of the astral body, according to Steiner, is desire. When I get even more personalized urges, I am in the realm of desire, and very often I need some external thing to bump up against my desire so that I know there are laws about this. The quality of action in the soul's will is desire, and the quality of that desire feeling is jurisprudence. It is a kind of *"Über-tact,"* amplified tact because it involves a whole other level of the soul—more than just concern about winning or losing. Now we are moving into the realm of animal nature. A cow might think that once the farmer leaves she can get into the molasses and alfalfa bin. This goes beyond the cow's life body concerns and into her soul life.

There are certain laws in nature that govern animal relationships. Too many predators put themselves out of business by using up the population they need for food. The jurisprudence of the astral body is reflected in that type of natural laws. We could say, then, that the base chakra, the chakra at the navel, and now the spleen chakra represent the forces of nature in the human organism that contribute to the spiritual challenges of human evolution. Instinct, urge, and desire are the lower octaves of the will forces that human beings eventually have to transform to become all that we can become spiritually.

In the ancient schools, students would plunge deeply into their lower chakras to meet the spiritual representative of those earth forces in the body. That spiritual being was a serpent coiled around the base chakra. The name of the serpent is *kundalini*. The ancients would meditatively enter the lower chakras to stimulate a circulation intended to bring the win-at-all-costs serpent–nature force up and into wakeful consciousness. The whole world then was based on tribal consciousness; if you are from my tribe, you are my ally, If you are from another tribe, you are my enemy. Transformation of this requires esoteric training that lifts the energies from the lower chakras into the higher ones. There were all kinds of techniques developed to do this, and over time human consciousness has risen from the root and is moving toward the crown. The deed of Christ on Golgotha was intended to support this process. Because of that act of sacrifice, humanity as a whole is now capable of working on the heart chakra as the new center for thinking.

The heart chakra acts as a kind of fulcrum. This is a good image to work with as we approach the idea of the heart becoming a sense organ and the eventual basis for a new kind of thinking. Biologically, the heart has the circulatory motif of every organ in the body. The motif of venous blood *in* is found in the liver; that motif exists in the right side of the heart. Venous blood *in* and arterial blood *out* is the lung circulation; it is found in the two ventricles. Arterial blood *in* and arterial blood *out* constitutes the circulation motif of the kidney; that is the circulation of the left side of the heart. Thus, the heart has the circulatory motifs of the other life organs—liver, kidney, and lung. As such, the heart serves as the central organ of sense and perception in the vascular system. It senses how the "I"-organization lives in the warmth in the circulatory patterns, because that warmth, the enthusiasm, is what moves the blood.

We will put the heart in the center, since I want to return to it. This is because the purpose of modern initiation is not to go down and talk to the snake; the snake has evolved; the reptile has changed into a bird; the snake has changed into the Holy Spirit. For contemporary initiation we do not go down and talk to the snake to get it to come up; rather, we say thank you to the snake. We want it to keep working on the unconscious part of my body down there. The central task of contemporary initiation is to develop capacities in the chakras that bring the Holy Spirit down. To do that, we have to work on the upper chakras, and they are *specula*—mirrors of the lower ones. This is what I want to share with you here. The quality of will in the chakras, including the biology of the glands linked to them, are pictures of how the lower faculties of instinct, urge, and desire manifest in the higher chakras. These behavior patterns manifest in the physiology as glandular secretions governing everything from reproduction to sleep-and-wake cycles.

We will leave the heart there as a kind of center (fig. 27). Drawing a circle through the throat chakra with the heart chakra as the center, the circle will also include the spleen chakra. A circle can be drawn that has the heart as the center and includes the throat and the spleen chakras. This shows a way of meditating on how desire can be transformed into a higher force called *manas*. Desire is a level of will in the astral body, and Rudolf Steiner calls the higher level of desire *manas*, which is the transformed astral body, or transformed desire. The code word for that transformed body is *Sophia*, or *Virgin Sophia*. The purpose of having desire in my astral body in the lower chakras is to serve as a kind of foundation for desiring, or wishing, to do better. My desire for things in the lower spleen chakra become the desire to do better in the manas body. The desire to do better means having to overcome my astrality. In Steiner's work, he connects the throat chakra to clairvoyant perception of the astral realm, to the zodiac and the planets in front of the zodiac, planetary movements. So the throat chakra has this kind of multiple function; it connects the lower parts of the soul to its higher parts.

Looking again at figure 27, we can see that the quality of the throat chakra is to perceive the astral quality of forms. In my throat I speak, and when I do I am actually creating astral forms.* This is what you do in eurythmy. In my throat I take lower forces, lift them, and create forms. It is a kind of reproductive cycle. Rudolf Steiner then connects the larynx and throat to the future Jupiter condition of Earth, when it will be an organ of reproduction, a kind of uterus. When Sophia is transformed, ordinary desire becomes the wish to do better, and I will be given the ability to create the world out of the *Logos* nature.

Logos potential can be awakened in my throat. Every time I pay attention to things that are word-like in the world—that is, the meaning of things that I take for granted such as trees and flowers—I begin to transform desire into manas. Normally, my sensory experience is just looking at a cow and thinking it is a cow. However, as I start to work on myself, I begin to perceive that even in the word "c-o-w" there are forces in the movements, gestures, and forms of the cow that represent laws of the astral plane. My throat is the organ of perception of those laws, because it gets its energy from my spleen chakra that has to do with digestive forces. In my throat chakra, I have to digest the form forces of the world that grab me by the throat, so to speak. These are the Logos forces of the speaking world. They are the forces of how the world's forms are formed. On the right side of figure 27, we see the astral realm, animal archetypes, and planetary motion, but then we have the laws of form and, as Steiner puts it, how others think.

How do others create pictures and forms that I perceive? When they think, they take the life forces of the life body and with their "I"-organizations make pictures and images. They make corpses of the organizing imaginations of the life body; in other words, they make thoughts.

---

* See, for example, Serge Maintier, *Speech – Invisible Creation in the Air: Vortices and the Enigma of Speech Sounds*.

Making thoughts out of imaginative potential is a type of reproduction. It is a kind of metabolizing force, as we have seen; making thoughts as images is catabolic. The astral body takes limitless things and makes them into words, meanings, and pictures as a kind of digestion of the world, leading to reproduction of the world in words and ideas. When someone is thinking or expressing something, the throat chakra is the place where speech organs are making the eurythmy gestures of the sounds they are speaking or singing.

At first, "throat perception" is very unconscious, but if I work with images, thinking about speech, penetrate gestures of speech and word-like things with form—if I model in clay the form of the embryo, or move in my limbs the consonants, *T, S, R, M, A(h)*, my throat chakra is being educated in the lawfulness of the formative forces. My throat is being developed into an organ of perception of the lawfulness of the forms of the world, the word-like forms of the world. Because it is the octave of the spleen chakra, when the throat chakra becomes developed, it becomes the tool I need to go down into my desire body and transform it. The throat chakra where I work with pictures in the astral realm—lawful pictures such as the embryo of the astral realm, convex–concave, and so on. I take them in, and even if my hands are moving them (i.e., I'm not speaking them) I am educating myself on the lawfulness of the form through the movements of my hands. My throat chakra determines whether a form is lawful or not. If the curve is filled a little with astral fat, I plane it off. That is training in the lawfulness of the forms in the astral realm. My throat chakra gets a step closer to being an organ of perception for my own astrality, because the throat and spleen chakras are linked harmonically.

What I am saying right now is really the goal of this book. Much of the rest of this information can be found in physiology books, but your physiology is a primer for your gradual evolution to sainthood and becoming a player spiritually in the world. Your own physiology is the book, especially the more dynamic aspects of the physiology that you find in the glands, because the glands represent the interface between the astral body and the life body. The great work is to understand this and work with it consciously. Alchemists call this the great work.

I have to take all this fluid stuff—remnants of what was once the content of the world—into my spleen and liver and transform that into my DNA. My organism takes the life force and, using remnants of the world, starts to create building blocks. Your spleen, gallbladder and pancreas are pretty much catabolic—the astral function is to kill the fat, kill the blood. Gallbladder, kill the fat; pancreas, kill the carbohydrates; spleen, kill the weak cells in the blood. This whole area has the hallmark of the astral realm. We could say that the desire of the gallbladder is to kill the fat. If you have in you a pattern of beliefs that it is not right to kill things, you will begin to have issues around the fat in your blood that can't be resolved; you can't pull the trigger for your Mars organ. When I want to go in and heal that, I have to understand what the desire in that space astrally is, and then see if I can find an analog in the higher, more evolved astral in my throat, which has the forms of the world that wind up being fat, sugar, and so on. Where do I find that? Well, it may be a tree, it may be a particular kind of flower, or it may be an animal substance or mineral. It has a particular form with which I work, because I recognize the synergy between that form and the dysfunction. At this point, I am moving toward the next chapter, on homeopathy and constitution.

To reiterate, the throat works with the astral forms of the world, so to speak. If I begin to penetrate the astral forms with consciousness, I start to see the various forms of desire in each them. I see the cow's desire in the little protrusions that come from its tongue and line its throat and go all the way down into the digestive area. On the tongue, all these little fingers sticking up on it secrete mucus so that the cow becomes a kind of mucus volcano. Each organ in the body is an image of a desire for something. If we follow the forms of those desire forms, we see a harmony of forms that integrate into the life and consciousness of that organism. The various organs evolve into organ systems, but they share a motif of the desire body of the organism for a particular sensory motor stimulus. We call this *physiology*. It is remarkably organized astrally to take the world and create a new being from it.

In my throat, physiologically I have a little mini-digestive processes going on well before the contents of the world reach my stomach. My thyroid is in my throat, and if we Google "thyroid function," we find that the thyroid governs metabolism. It governs metabolism, because the issue is the astral form principles in metabolism. My pancreas has to disassemble, and my gallbladder has to dissolve the forms of the world into discrete simple substances. My stomach has to make acids so that I can break down complex carbohydrates. Lack of the correct acids in the stomach is the beginning of the diabetic pattern. My whole digestive organism has to deal with the forms of the world. In my throat, I deal with the forms of the world, and the whole desire principle revolves around what is word-like in the world.

Between the throat chakra and the spleen chakra is my heart chakra, where my "I"-organization lives in the warmth that allows me to make those transformations. The heart chakra focuses on determining what in my life is progressive and can lead to development and what, by contrast, is retarded and will no longer lead to further development. To do this, I have to perceive, in the intent of my actions, *why* I am acting in the first place. Steiner calls will at the level of the "I" *motive,* because the function of will in the "I" is to perceive the motive for any particular action. In the soul or astral body, the will is based on jurisprudence. At this level, the motive for changing an intent in my will comes to me from outside my soul in the realm of "laws," which is jurisprudence.

In my "I"-being, my ability to see the motives behind my actions allows the desiring will in the soul to be transformed into manas, or the wish to do better. Biologically, however, this means I have to be able to breakdown and digest the world I have constructed out of desire. This is all your therapist will tell you: Try to look at your life in a new way. What if it was this way? How do you feel now? This means metabolizing, digesting, and dissolving desire as the motivating force in one's life to create manas. Then what happens? When manas arises, there is a new meaning of the same situation. This is the function of the throat chakra. It enables us to look into the astral realm that governs and is connected to the whole metabolic realm. What I am saying is that there are octaves of relationships between the biology of the organism and the soul element of the organism. When these are harmonized through inner practices, our body and soul can be used to create sense organs of the spirit embryo, to which we will give birth in the far future.

Now, moving upward from the throat chakra we find the brow chakra. You can draw a curved line from the brow chakra, where it says "integrates senses and thinking," down to "power / tact," or urge and prudence. Rudolf Steiner gives a picture of how the brow chakra allows us to integrate sensory activity with thinking. This union leads to perception of the etheric realm and plant archetypes. He calls that etheric realm the *lower devachan*. This is code for the etheric realm, in which plant archetypes live as the basis for the etheric, life body, or life forces. The brow chakra at a higher level of clairvoyance allows us to penetrate mysteries of the etheric.

The brow chakra is primarily the site of our pituitary gland, but also includes the pineal. We see in the brow chakra the higher etheric realm where the archetypes of life live. If I transform part of my life body by truly changing my diet, I could get my ether body to come more in line with the archetype of life forces. By changing my diet, for example, I create a force in myself for understanding life as a force. Steiner calls that life force in my soul *buddhi,* or buddha life. The buddha issue is this: Why do we come here to suffer? The mission of the Buddha is to overcome the tendency of life to be destroyed by desire. As an initiate, Gautama Buddha brought into the world a strong example of a nascent quality in our souls called *buddhi*. It is created in the higher dimensions of our soul life when we overcome a desire. When that happens the life body is lifted into consciousness. It is no longer filled with desire, urge, or drive that pulls our life forces down. The ability to control desire becomes what Steiner calls *resolve,* or *mother-love.* Resolve is a high level of will that is clear of self-serving intent. It is not the highest level of will, because mother-love includes a hidden feeling of needing reciprocation. The highest level of will is True Love, which is unconditional; there is no expectation of reciprocation or attachment to an outcome in response to performing a deed. Mother-love is a level below True Love. It is present even in the animal realms as resolve to self-sacrifice for the good of the offspring. That force is the urge of the life body in the tribal consciousness taken way up to a high plane of self-sacrifice. When a mother animal sacrifices herself, she gives the substance of her body to the offspring during gestation. This is the great sacrifice of the body and life forces. If that is the guiding principle in the soul in general—if the soul is working on buddhi—then one's life is lived for others. That is Buddha, the sacrifice of one's life for all sentient beings.

Women will tell you there is a life body connection for about a year and a half after childbirth, and then the child cuts off that connection. Up to about a year and a half there is an inner bond between mother and child through "strands" of the shared life forces. Then the child cuts it off, and there is a deep feeling of loss. It takes a while to feel the separation, because the woman has given away her own body, and then a part of that body walks away and becomes interested in things other than the mother–child relationship. It's little feeling of loss. This condition takes away the unconditional aspect of the will nature. Unconditional will is being glad when someone leaves and finds wings to go. This can happen, but it is not the usual dimension of mother-love, especially early on. True Love can grow out of mother-love, but to make the transition is tough. However, the great cosmic playbook says that as human beings we will eventually reach unconditional love.

Looking at the function of the pituitary gland behind the brow chakra, according to science, a primary function of that gland is to regulate growth and reproduction. One of the main ways it does this is by sending out hormones that regulate the way water moves in and out of things; that is how life forces act in the body. Again, everything around the brow chakra points to the realm of life that Steiner calls lower devachan, the realm of plant life, the realm of the life body, in which the forces of life come in from the starry realms. The soul aspect of this is to integrate sensory motor responses with cognition. The brow chakra and pituitary gland are involved in the effort to connect something through a sensory experience with a thought. When I do, I gave birth to the thought.

According to alchemists, thinking is a higher level of reproduction, which is a higher level of the excretion. We all know what happens in excretion and secretion; something falls, and levity forces rise. What comes up when something falls out of the pituitary gland is thinking. The "I"-organization uses forces of thinking, the forces of life, that arise from the deposition in the secretion to create inner pictures, or little reproductive images. Men, who have no womb, use their head as a womb to birth ideas that they then treat as surrogate children. This is a form of jealousy of men toward the creative genius of women. All of this has to do with etheric forces. For a man, the etheric forces make a womb in the head and the life forces become thinking. For a woman, however, that kind of thinking can actually descend all the way into the uterus and become world thinking, creating images that become individuals who get up and do things in the world.

The brow chakra can lift us into the lower devachan and connection with the forces of life. If I work meditatively on connecting sensory experience and thinking (which is the whole purpose of this book), an organ of perception grows in my brow to work with pictures of things as they appear to me as fixed in the world and to morph one picture into another inwardly. When I do this, I see how the ether forces link the separate images. I can learn to look for the pregnant space between musical notes, where the true music lives but is not manifest. The throat sees proper form, but the brow sees *how the dynamic form becomes a fixed form*. I have to link a sensory perception to the correct inner picture and be able to keep those two moving through the sequences of pictures—the movies in the head that we call thinking. When I work with this chakra and develop a higher faculty there—when I correct my thinking and bring my thinking in contact with lawful pictures of the world, I form pictures of the world, dissolve them, and then keep working with them across the threshold until they correct themselves—I am creating embryo buddhi in my spirit. I am creating a life body for myself that I will be able to use in later stages of evolution. I harmonize my will with the creative force.

Leaving the brow, we go to the crown chakra. We can draw a circle from the crown chakra all the way down to the chakra of instinct. Rudolf Steiner suggests that we intuit high states of consciousness in the crown chakra. The highest states of consciousness are those that are eternally true. He has a code word for this that came from Theosophy—*higher*, or *upper, devachan.* The ancients called this *the crystal heavens,* the realm in which minerals have "I"-being. Within that realm is the great lawfulness of the hierarchies' eternal imagination and their activity in creating a world that we, the tenth hierarchy, will eventually make as lawful as the laws that govern crystal formations.

This is the great secret of the alchemical stone. In the end times the Earth will be a harmonious and geometrically lawful world, because then we will have an organ in our spirit embryo called *atma,* the completely transformed patterns of instinct that govern the physical body. In atma-consciousness, these physical patterns are lifted away from "winning" to becoming the "I Am." In that consciousness we experience our "I"-being embedded in the great "I Am" with everyone else. That is atma. It is a kind of mineral consciousness, because there is no vague, sloppy stuff. We look at each other and understand what needs to be done; we agree to it; it gets done; we move on—competence, autonomy, joy. This is the way it is in the world of minerals. There is creativity and flexibility connected primarily with crystal forms. The various crystal forms are related to very minute differences in temperature, pressure, salinity, and viscosity. This is a picture of the "I"-being activity there.

With the heart chakra, it's like we've gone up to the mineral. In mineral consciousness I am still in the hierarchies. I am in the creative world, but human beings have a seed in their heart to go beyond the hierarchies that create nature, to become creators of a new cosmos. We do so because a member of the Trinity has come into a body of flesh and lived among us human beings. Not a member of the hierarchies per se, but of the Trinity, the ineffable beyond the beyond. The great wisdom has taken the forces of the sense body and preserved them, so when that being came into a body of flesh, those forces could be returned to humanity as a seed. This is Christ-consciousness.

There is an even further realm of freedom—will that involves the highest level of free will, which Rudolf Steiner calls *love*. As he puts it, when I will my thinking, that is freedom; it is meditation, concentration, exercises, and so on. When I will my thinking, that is freedom. And when I then think into my will and penetrate my will, I see the motives for my thinking, and that creates the seed of love in me. Steiner characterizes love as becoming so vulnerable that I allow someone else to see who I actually am, which is a scary thing. The organ that will hold that capacity (we've run out of organs going up to the top) is the heart. The heart, as a sense organ, includes and incorporates all of the organs and takes it a step higher, because it contains the biological motifs of all the other organs.

The moral and the soul and the biology are interwoven, and the practices we do working with pictures and the arts—with eurythmy, sound, color, cooking, growing food, paying attention to cows in a loving way and to the sacrifice of the beings that are below us—all of this is actually training to create this spirit embryo through this kind of step-up or step-down transformation of the resonance between the biology and the soul element in the chakra and the spiritual aspects of the soul forces, and we finally get the transformation to happen. The key is in the heart and the way it relates to the pituitary and pineal glands through the door of sleep, taking pictures that we work out in the world into sleep and then retrieving them meditatively in the morning. This is the Rosicrucian stream.

All of this spatial–temporal stuff makes it possible to telescope our time and take advantage of our space, which means working with pictures by taking them out of space and time and moving them through ether sphere up into lower devachan, upper devachan, and beyond, and up into the Christ sphere, where the spirit embryo lives in these higher vibrations. Our body and its forces are the primer for this work. It is the owner's manual. The patterns put there by the hierarchies are what Steiner calls the petals of the flowers of the chakras, saying that they are only half of them; this is our body and all the beautiful forces and organizations in it as it exists as half of the chakra flower.

Our task as human beings is to build the other half of the chakra flower. The other half will transcend the forces of the first half, but must be built in harmony with them. That is the caveat. The fact that we have free will is the difficulty. We can build a whole new edifice that is not in line with those original creating forces, and then we need the great stick of the hierarchies—like the stick we use when herding the ducks—and that stick is illness and pain. It is the stick the hierarchies use to get us to go where we should be. The chakras are the blueprint of this whole process. This is why the chakras are so important and why the Rosicrucian imaginative faculties are so important. Our physiology is the source of the imaginations that need to be lifted. It is the source of the creative force. It is why Rudolf Steiner said that, if more people study physiology, the social forces in the Anthroposophical Society will greatly improve.

Chapter 12

Remedies and Dysfunctions

"The Christian–Rosicrucian method of training human feeling and the will involves transforming the soul through imaginative pictures to cleanse the astral body of its reliance on inner pictures that depend on sensory perception." Instead, the student must be illuminated by the imaginations that exist in the cosmos. Cosmic imaginations need to be given back. There are ways to do that. In the patterns of things we can find the way to a Rosicrucian technique. "We must first build organs of perception for those imaginations in fully awake consciousness so that during sleep the astral body can imprint its experiences on the life body. [Rudolf Steiner calls this process *catharsis*.] The portion of the astral body illuminated by the cosmic imaginations is called the Virgin Sophia. When it approaches the cosmic 'I,' its light shines down and is received by the Virgin Sophia. We call that illuminating universal 'I' the Holy Spirit."* People with these capacities lose their tendency to form opinions. The self is eclipsed and they communicate cosmic wisdom rather than personal views. More important, those individuals speak without sympathy or antipathy toward any particular worldview. This is a very important aspect of Christian–Rosicrucian initiation.

Rudolf Steiner has located in that realm of the cosmic ether what he calls the *life ether*, which is life abundant. It is a realm of continual life, the farthest reaches of the cosmos. That continual life is centered in the infinite. But because there is a consolidating force and a substance, the substance has a center that interacts with the forces of life at the infinitely distant and

---

* Steiner, *The Gospel of St. John*, lect. 11.

creates a kind of field of tension in between. In projective geometry, it is called the action of the point at a relative center and the forces of the infinitely distant plane experienced as a circle of infinite circumference.

We have light and matter as the two polarities. The source of light that casts a shadow is the light that does not cast a shadow. That light is universal, Christ consciousness; total light but no shadow. As that light moves toward a relative center, as it did when the Word became flesh, there is a wake in the passing. As Rudolf Steiner puts it, there is a life ether composed of all of the ethers below it. Then, falling out of that life ether is a chemical, or number, ether, or the music ether. The musical ether is the pattern in which life arranges itself to create potential. It is the music of the spheres, where nothing is manifest but the potential of the whole arrangement is held as a whole symphony of possibilities. The potential for the patterns of all substances is highly organized in this harmonic ether, but nothing is manifest. Then, out of that chemical ether, the light ether falls and comes closer to a state of manifestation.

This is the light that just begins to be able to create shadow. The light ether supplies the energies for the chemical ether. Gamma rays, ultraviolet rays, and visible-spectrum light interact with oxygen molecules and other gasses in the stratosphere to create ozone, energizing oxygen production that keeps oxygen in the atmosphere in a constant balance through processes of dissolution and regeneration. The areas for this are in deep space, where the chemosphere and thermosphere provide protection for all life on Earth by absorbing the ultraviolet rays of the Sun. Finally, the lowest of the ethers is the warmth ether. In the stratosphere the interaction of oxygen and UV radiation releases warmth above the Earth. On the surface and in plants, these reactions between light and the chemistry of plant pigments create photosynthesis, which creates carbohydrates for our food and makes oxygen for our life.

According to Rudolf Steiner, in our soul life, the warmth in the blood just happens to be where the ego and its "I"-organization live. The shadow of that warmth, the corpse of that warmth, is known to alchemists as *fire*, the shadow of warmth. The pattern for this is the element as a shadow of an ether. The ether is an energetic template, and the corpse falls out of the ether and manifests as an element.

We are now out of the ethers and down in the elements. The corpse of the light ether, when it falls, becomes the air element. The corpse of chemistry, number, and music is the great organizing chemistry of water. Finally, the corpse of life is the mineral, the basis of consolidation forces. Life is the animating ether that contains all of the others. Water as element is the basis of the personal etheric. Air as element is the basis of the astral organization. Warmth and fire are the bases of the "I"-organization. The falling of the ethers into elements eventually ends, through consolidation, in the formation of substance.

Substance takes particular pathways through various ethers and elements, accentuating one or another, depending on which zodiacal sign the starlight comes from, which planet it connects with on the way down, and so on. We could call it the line of emergence of substance. The form of substance and its molecular relationships seen esoterically is a picture showing the wake of its passage through this step-down process. The palette of all the various substances is the harmonic scale of the spheres. Science uses the *periodic table* to describe this. The fall of substance into matter reflects the action of how the physical body makes its structures based on forces that consolidate into a form. The mantra for this is "form is motion come to rest." All these movements come down through the energetic form, or pathway, into manifestation, and then the substance occupies the manifest form that is an image of its path into the visible world. As a sense-perceptible experience, the process and the form carry the biography, or ontogeny, of their passage, the evolution of the one, the being.

In substance is a kind of rubric, encyclopedia, or reference manual for how processes die into the corpse of matter. This is the nature of substance from an esoteric viewpoint. If I work with these images and concepts inwardly, the form of a substance will reveal the dynamic of the substance. The form reveals its action, or dynamic, in the natural order. We might call this pharmacology. When I give a particular substance to a body, especially a human body, that body has a whole, huge palette of all of these possibilities to recognize the substance and determine whether it needs that substance or not. This is because the human being is the archetype for all of this; every step down in the process of an ether moving toward substance is an extrusion down and (by now a familiar refrain) there is a release in levity force for the purpose of consciousness.

The diagram on the next page shows twelve positions that represent characterological dispositions. They describe the conditions through which human beings, when incarnating, pick and choose planets and star groups to create an "I"-organization—not the "I" but the "I"-organization, the part of one's "I" that will have to deal with the physical body. As my spirit comes through the spheres, I pick and choose directions, patterns, and movements that become consolidated into my tissues on my birthdate. The forces I have chosen in incarnating become imprinted into me, and we call that imprint a brain. In technology, an icon symbolizes a whole set of relationships that allow you to do things. Your icon is your brain; the app to which it gives you access is the karmic plan of your life. That is your brain. It is a kind of a code for the forces that you want to use and apply in living your life. Karma is the plan of your journey toward manifestation through the web of forces composed of the cosmic etheric and astral and the light ether, chemistry, and elements you pick up along the way into incarnation. These force fields become the template of your body and its processes, the way the organs will be arranged, the ease with which your liver can communicate with your gallbladder and the quality of the energy that you have.

The spiritual human being descends step-wise into physical matter. At the highest level, my True Self, or individuality, is embedded holographically in the Christ-being, the being of the Cosmic I AM. I am a drop of the ocean of being, with Christ as my benefactor because the Christ-being experienced the separation of individuality in a physical body. The True Self, or "I"-being, is a state of consciousness in which my individuality is separated out of the totality of the I AM but still carries the vision of the I AM; it's still in contact with the I AM. Then lower

Rosicrucian Esotericism

Why is it that when sleeping, human beings are unaware of sensory impressions although they are surrounded by sensory objects? During sleep the physical and life bodies of humans remain in the bed. The astral body and ego emerge and are in the spiritual world. Why then do the sleeping human astral and ego bodies not perceive the spiritual world that they are in? This is because the astral body of the average human has no organs of cognition for the spirit. Through initiation the astral body is provided with spiritual sense organs that are capable of following the spiritual experiences that are undergone in the night. To do this the inner life of humans needs to be strengthened. In the student's feelings and thoughts there must be a conscious effort to focus attention on subjects that relate only slightly with physical reality or we could say sensory unreality. Mental pictures that represent remembered images of sense objects are not suitable for entering into these realms. Contemplating symbolic pictures that are linked to particular feelings and attitudes of soul are the most effective means to building the astral organs. Abstractions are not effective in that they lack the feelings and perceptive experiences that link the inner work to the astral body. There are three paths that can be taken towards the development of these astral organs of perception. The eastern path is the path of devotion or surrender. The Western Christian path is the path of mystical union and piety. The Rosicrucian path is the path of thinking and will training.

---

staphysagria
+creative, inspired, charismatic, committed
– indignant, incised, wounded, fears the practical life, martyr, self destructive –

lycopodium
+tactful, benevolent, tolerant, charming, magnanimous
– a free mind with a fixed soul, habitual denial, aversion to change, supports underdog but undermines equals

pulsatilla
+ devoted peace maker, sees beauty in everything sacrifices self
– easily crushed, fears anger, needs noticing, guilty conscience when angry,

sepia
+ transcendent soul life, deep spiritual roots, devotion to causes especially in the arts, self reliant, candid
– estranged from life, avoids affection, soul sore, complaining,

sulfur
+ thorough, prolific, learned, direct
– self centered, caustic, overwhelming, inflated

ignatia
+highly intuitive, perseveres through obstacles, patient
–exaggerates inner drama, grieves unknowingly, self absorbed, numb to life

arsenicum
+ very enthusiastic, loyal, organized, perfectionist
–, driven, domineering, nervous, excitable – the general, the throughbred.

calc carb
+ mystical, intuitive, loyal, deep self knowledge, true believer, supports causes
– stubborn, brooding, afraid to act, fantasist, lives in the past

phosphorus
+witty, high energy, learns easily, many ideas
– scattered, absent minded, confused about identity, multiple personalities –

natrum mur
+truth seeker, keen intellect, fair, just
–opinionated, bitter, relentless, loner, absolutist, unapologetic

silica
+sensitive, rich inner life, capable, self limiting, meticulous, practical
– apprehensive, anxiety attacks, seeks recognition but fears being seen

lachesis–nux
+ competent, high ideals, diligent, organized, insightful
–spiteful when betrayed, aggressive to others, closet anarchist, irritable, conflicted over morals especially sex

than that level is my "I" that falls out of my "I"-being as my individuality when I realize authentically who I am as an "I," and I say "I." This is a rare state of awareness, but it is a peak experience that gives me access to Christ-consciousness and becomes a life-changing event. Lower than that is my "I"-organization, which most of the time, when I say "I" really refers to experiencing my "I"-organization operating in my body. Then lower still is the experience of my individuality as the Great Me. That is the lower self, or ego as defined by psychology, which is the source of my personality and supports itself with self-feelings, unchecked impulses, desires, and opinions. The core inside that is a flow of deep feelings in my body related to its urges and instincts.

Instincts, urges, and, desires form a personality that has a particular characterological disposition that is automatic until I develop the ability to see the fruit of my will impulse before I enact it. This depends on how much I have been able to relate to my "I"-being as a unique spiritual individuality. There is a hidden, seemingly paradoxical law in this. The more individualized I become, the clearer I become, the more my consciousness can expand to a perception of the universal. If you play for me now three notes of a particular fugue of all the sets of art of the fugue, I could most likely recognize that it was music of Bach. A really good music person would know which piece it was from after hearing just three notes. This is because the more individualized Bach became, the more he harmonized with the great central core of the universal human through his "I"-being. There is a kind of flip that happens. It's ironic that the closer you get to penetrating the mysteries of your own body—a Rosicrucian task using science to go into your own organism—the more you start to experience how it would be to experience deeply the light of consciousness falling into matter.

Some people specialize in the consciousness of the light falling into matter. They wake up in the morning and are aware that part of them is an eternal being, but a part of them has also fallen into matter. Being aware of that fall somehow becomes a motif in their life. They wake up in that consciousness and the impulse they feel is to serve humanity, and they watch during the day as the light of consciousness falls into matter. We could call that type of person a therapist. They are sensitive, and they watch. Such a person could be a caseworker or a social worker who works with indigent or homeless people. They watch how light falls, so they carry with them the perception that everywhere in the world where light is falling into matter they wish to be of service. Eventually, this creates in them a sclerotic tendency because, day after day, they are sensitive to the fall of light.

Say, for example, that at the motor vehicle department is a woman who shows up day after day and works under the fluorescent lights, taking pictures for drivers' licenses. She got this job with the government because she grew up in a family of activists who believe that working in the political realm and government are important civic duties and one of the best ways to serve the world. However, she pushes a button all day and, although a part of her is serving, another part is sensitive to the "fallen" condition of it all. Over time she develops a skin condition, because she needs protection in the form of some type of sheath. She develops psoriasis on the elbows, or warts or moles. She gets these deposits on the surface of her body because she is so sensitive that her wish to serve manifests in a seemingly senseless way.

The woman is trying to cover herself but feels that her ability to do this is compromised. She lacks the right kind of consciousness to protect herself, but she cannot tell herself to quit this dead-end job and go make coconut dolls in Central America or some such thing. She really wants to serve, so when she starts to develop skin problems she needs to find a substance in the world that is an image of her dilemma—a substance that is serving everyone and is fixed in a niche where the light falls into deadness as a result of its service. Here, it would be a good idea to look at the mineral world, where the highest consciousness gets caught in the most fixed conditions. When I look at a geology book for the mineral that is the source of most other minerals, I find silica. Silica becomes many other minerals; it serves the world by sacrificing itself to become a wide range of silicates through weathering and breakdown.

When I take silica, grind it up, shake it in water, and then take a little dose of it, this would be a homeopathic remedy for skin problems—skin that has become hardened and formed little crystalline forms or warts that look like little crystals sticking out of your arm. That is homeopathic silica. The soul mood of silica, as a substance coming down from the life either into a mineral, creates a kind of pattern, and then even in the Earth's body it serves that pattern.

Taking that substance and lifting it from substantiality back into its activity by diluting it rhythmically, I get a remedy that works well for the sensitive server or silica constitution. I move the pulverized substance rhythmically in water and dilute it incrementally in steps to render it into less and less of a substance and more and more of an activity. As I do this, I start to move the substance up into its soul quality, or air body. Moreover, if I take it way out and attenuate it out into almost nothing, I have moved it up into the light and chemistry of the life of silica, where silica is a living being rather than a corpse.

The purpose of homeopathic modalities is to lift a substance back into its active being. The task of the homeopathic doctor is to diagnose the harmony between the path that the substance takes to come into manifestation and the soul pattern that the person is manifesting in life. That harmony is *characterological disposition*. When we take a remedy, our life body, soul body, and astral body reconnect with our "I"-organization in an archetypical way that begins to heal stuck life forces.

I may have created a certain disposition that I have to overcome—for example skin problems. It may be that I have skin problems because I really don't want to be looked at, perhaps because in a previous life I did all kinds of crazy things to gain attention. Maybe I was a harlot or in a harem. To balance this, I came into a body in which I will have genital warts. I choose certain patterns when I come in to have a disposition, but not just a random disposition, because I could choose arthritis, which would have nothing to do with my genital area. If that were the case, I would choose a different remedy to ameliorate the same thing, depending on how my character does what it does. This is where the soul, the substance, and the rhythm of the remedy interplay, because I have to take the homeopathic remedy in a rhythm, and it is prepared in a rhythm, and the substance is actually lifted away from its substantiality into its imagination of becoming. Then my soul and spirit come into contact with the imagination of becoming, and very dimly I begin to remember why I needed this. If you can remember why you needed a dysfunction, it is the first step toward not having it. That is the key.

Some therapists will tell you that they can give you all kinds of remedies, but until you actually decide you no longer need that illness, they are just dumping useless remedies into the body. They can give you a new heart, and in a year it is all clogged up again. They will give you another new one, and in a year it's clogged again, until you understand why it gets clogged. Then they give you a new heart and it is fine, except for the fact that you take on the personality of the person who had it before. Why the heart? If you get a liver from somebody, and it doesn't happen. But in case after case, a new heart is given and the person suddenly becomes a completely different person. He may be a fifty-six-year-old guy who had erectile dysfunction for thirty years, and then he gets the heart of the twenty-nine-year-old Puerto Rican guy who was a rock star. When he comes home, his wife is very pleasantly surprised.

Homeopathic remedies and flower essences are forces in the suprasensory realm of the ethers, and the elements that result in a remedy that allow the soul to connect with pictures of why things are happening.* This is very different from pathology and pharmacology aimed at eliminating organisms, because one gets rid of organisms while the state of the organism is unchanged. With flower essences and homeopathic remedies, the goal is to tell the "I"-organization about how the body is operating so the body can do the healing, since it is understood that the life body is never ill. The problem is just a miscommunication between the "I"-organization and the astral body, based on the work of Lucifer to create a condition of illness.

---

* See, for example, Julian Bernard, *Bach Flower Remedies Form and Function;* and David Dalton, *Stars of the Meadow: Medicinal Herbs as Flower Essences.*

If I can take a substance that is an image that reminds me of how this process developed, then my ether body sees why this is. Then I have an "organ dream," which occurs when we take something into sleep, especially something that has a feeling connected to it, that allows the astral body and the ether body to remain very subtly connected, even after they separate. There is a very delicate connection, which is in what Steiner calls *the sentient body,* the lowest part of the astral body, or soul body that serves as an agent for linking to the activity of the "I"-organization. It is a repository of feelings of how things are going. The sentient body is the interface between the soul and the life body. If I can place pictures into sleep, I begin to gain access to very healing imaginations.

Now suppose I find that I have a silica constitution, for example. We see this at the bottom of the diagram in this chapter. The positive aspect of my silica constitution is that I am sensitive; I have a rich inner life and feel very capable, priding myself on seeing things clearly. I have the quality of being self-limiting, because there is a certain kind of mineral quality to my consciousness. I like things tidy, because I have a strong sense of the structure of silica, where light falls into matter. I am self-limiting, meticulous, and practical.

Silica is the basis of computer technology. It is ubiquitous, sensitive, and holds memory very well. This is its consciousness. It is a kind of upper devachan visitor here, showing how it will be for us in the end times. I believe that is why Ahriman is using it to tweak our noses, because when we reach silica consciousness it will be silica as a cosmic consciousness of very high beings. It is the technological condition of silica that allows Ahriman to get involved. The positive is that I am sensitive.

The negative aspect of silica is that I become apprehensive because I am so sensitive. I recognize in some way when things are not harmonious—not conceptually, but I know for sure. I am sensitive, but then I walk into a particular space and know something is going on and become apprehensive and anxious, because I am aware in my sensitivity—especially on the surface of my skin—of any weirdness around me. As a result, such people are prone to anxiety attacks, because they know that the light is dying. They just know. They get up in the morning, and the light is dying, so they are just going to go to work and do a really good job, and nobody is going to get on their case. But they are always aware of where the next little thing is going to create turbulence. Their anxiety attacks lead to a deep feeling of avoiding being singled out. Although they may want to be recognized for their sacrifice, they also have a deep feeling of being revealed and have a strong desire to be very private, because they are silica and live in a glass house. They have a tendency to repeat things that they do, which is connected with a general anxiety, because they build glass walls around them that they can't see. Then, when they go to do something they smack into the invisible wall, because they forget it is there.

This is a picture of a substance related to serving and being tremendously sensitive. This is why our information technology is all about silica; it is sensitive. If I come in through that silica stream, I carry some of the consciousness of the great being of silica in the Earth's body. I follow that stream because I know that I need that for my life.

The opposite kind of consciousness (going across the circle of the diagram to *Staphysagria*) is a Delphinium flower, or Larkspur. The form of the Larkspur is a flower with a long nectary and a little ball at the bottom. When the insects come, they have to go all the way down into the flower to get anything; they have to penetrate it very strongly. The flower is purple and dark blue; it grows in the shade, under trees, and is extremely poisonous.

Poison in any plant is an activity that has been turned inward and cannot express itself outwardly. But when what has been turned inward is put into another organism, the poison aggressively takes over one of the systems of that organism. All poisons have this quality. Many poisons come from an incomplete protein-formation process. That symptomatology often forms the basis for alkaloid poisons and psychotropic substances. The life forces used in the formative process get turned inward and fester. What would normally result in a protein doesn't quite get there, and the incomplete activity creates a deposit.

When this activity enters another organism, the activity unfolds to become complete. It releases an incomplete consciousness that goes in search of something; it is "bandit" consciousness. It is not lawful; it is astral and yearning to be complete. It offers to other levels of consciousness an opportunity to collaborate and to be part of a higher organism. In this exchange, the higher organism gets to see the world from the view of the plant-being yearning to be complete. In a lawful world, the plant could not experience the perception of the higher organism, but when the incomplete protein process becomes incorporated into a higher organism the plant gets to experience completion of the process through the sensory motor and feeling capacities of the higher organism. The yearning for completion is satisfied and the plant poison uses the sugar in the blood of the higher organism to complete the metabolism.

You go down to the 7-Eleven at eleven o'clock at night and get three bags of Cheetos, and then you stop by Domino's for a pizza, because the poisonous intoxication of the plant-being has completed its metabolism with your blood. Toxins are created in this kind of radical catabolic process of poison formation, and then a radical consciousness arises from it, but now we are slipping into the realm of being.

I presented a snapshot of the silica-being, but now we are talking about the being of plants that seek an astral body to incorporate into their own organism, rather than just being a good plant like a carrot with an astral body that is in the spiritual world outside of the plant body. We give names to the plants that have an astral body inside—names such as *Mary Jane* (marijuana) and *Mescalito* (peyote, or little mescal), because those poisonous plants create drugs that interact with other levels of consciousness.

Staphysagria is a poison, and the brewing poison in this particular flower is a very deep interiorization of what is normally external to the flowering process. In the soul who follows the staphysagria track, the gesture is not one of serving or being sensitive; rather, the gesture is a deep interiorization of everything, to the point where the sensory world becomes almost like a poison. Out of that poisoning there is a reaction in this type of person that has to break out of the poisonous prison: I have to break the walls down get out of here. This is the polarity in the *Staphysagria*. The breakdown of the substance and creation of the poison, and then the breakout of that is polar to the invisible-wall building of the silica soul gesture.

Silica and *Staphysagria* are two poles. In *Staphysagria*, if the "I"-organization in this constitution can manage to hold on to the poisonous inward focus, breaking out of the poisonous prison is creative. There are those who couldn't care less about laws; they just go and create; they are always pushing the envelope. They continuously push against accepted forms in a lifetime of continuously breaking down. As a result, they are pathologically impractical and do really stupid things that, when they are creative, eventually turn to gold. However, when they are not creative, such people just live impulsively. This is the situation when we hear of someone snapping and harming someone else, and then blaming the victim. The constitutional feeling is one of being wounded because of everyone else.

The opposite side is *Silica:* I am going to serve everyone else, although it's a burden for me. In *Staphysagria:* I exist, therefore I am wounded, and I am going to let you know about it. Either I bend and break the laws in a creative way, or I bend and break the laws in an uncreative way by acting out my wound by projecting out on everyone else. Thus, the positive elements of *Staphysagria* are creative, inspired, charismatic, and committed to creative vision. These people are volcanoes of creativity, composers and architects.

According to an old story, Mozart went to visit Beethoven, and on the piano were seven plates with moldy chicken and half-eaten potatoes, suggesting that Beethoven was so creative that food was merely something to support his creative process. This sort of constitution has no inhibition about breaking rules if it serves a deep, inner, creative vision. The negative side of this is that this constitution is indignant over having been wounded; but creative force arises from the feeling of being wounded. There is a feeling of being incised by something that has since healed over into the unconscious sphere, but is nonetheless festering and creating the inner pressure that drives a creative urge. Such people fear the practical life, because if they actually ease the inner pressure, they fear that their creativity is going to go away. They live life as martyrs for the whole, and they can be very self-destructive.

These two characters are opposite to each other, and in life they often become attracted to each other. The flaming genius has a wife who helps him tie his shoes in the morning before he performs a concert, and she takes care of all the logistics for the concert tour. The *Staphysagria* lives in a deep inner world that always threatens to explode into another creation, and the practical, sensitive server likes to be associated with this explosive creativity, because it can be fun to be part of the explosiveness while remaining behind the scenes. In the *Staphysagria,* the mood is one of being entitled to adulation because of carrying the wound for everyone. Their characterological disposition is the noble sufferer, whereas the opposite is the sensitive sufferer.

I bring this polarity as a picture of how we can assess remedies and interactions of the soul with the substances in the world.

The unusual way of thinking about this is that, owing to the fundamental differences in consciousness, these two types will have very different organ pathologies with the same illness. This is where we have to look at more than the just illness and do a prognosis and analysis of the illness, and then go to, say, page 37 of the *Vade mecum* for the answer, but also ask how the soul processes in this person resulted in this health challenge. The homeopath might have completely different remedies for the same problem. In conventional, materialistic science, this is just a lot of mumbo-jumbo and does not make much sense, because the germ theory and antibiotics will work for everyone.

What I've tried to bring to you here is that this component of consciousness and soul force is the exact flip-side of the work in the organ systems in physiology. That is why this course is in *esoteric* physiology. Healing dysfunction is helped by a combination of the remedy and the soul remedy. If you take a substance to heal yourself, it is very useful to find a myth or story that reminds you of the manifestation process of the substance from which the remedy is derived. Along with the story, make little wax models of the different stages of the myth or paint a series of pictures that reminds you of the feelings of the fairytale characters; this will make the medicinal substance more effective.

This is why doctors in anthroposophic medicine will often include eurythmy, sculpture, painting, or speech in the therapeutic process, because they recognize that presenting illness as a soul constitution is equal to the actual development of the illness as a biological process. The sequence of feelings related to it have to be addressed for the medicine to work. The work done anthroposophically and based on the work of Rudolf Steiner uses all these modalities and even homeopathy and flower essences. But you can even give flower essences to a horse, and it works really well.

What is present in this medicine is the energetic imprint or template of the consciousness that came through the process of depositing. As those deposits were happening, consciousness arose, ending eventually as the great deposit we call our physical body. Within our physical body is an inflammatory process that is the beginning of healing.

Bibliography and Suggested Reading

Bernard, Julian. *Bach Flower Remedies Form and Function*. Great Barrington, MA: Lindisfarne Books, 2004.

Dalton, David. *Stars of the Meadow: Medicinal Herbs as Flower Essences*. Great Barrington, MA: Lindisfarne Books, 2005.

Frazer, Sir James George. *The Golden Bough: A Study in Magic and Religion*. Oxford, UK: Oxford University, 1994.

Holdrege, Craig. *The Dynamic Heart and Circulation*. Longmont, CO: AWSNA, 2004.

Husemann, Friedrich, and Otto Wolff. *The Anthroposophical Approach to Medicine: An Outline of a Spiritual Scientifically Oriented Medicine* (3 vols.). Hudson, NY: Anthroposophic Press, 1982, 1987, 1991.

Lehrs, Ernst. *Man or Matter: An Introduction to a Spiritual Understanding of Nature on the Basis of Goethe's Method of Training Observation and Thought* (3rd ed.). London: Rudolf Steiner Press, 2013.

Mantier, Serge. *Speech – Invisible Creation in the Air: Vortices and the Enigma of Speech Sounds*. Great Barrington, MA: SteinerBooks, 2016.

Pelikan, Wilhelm. *Healing Plants: Insights through Spiritual Science* (2 vols.). Chestnut Ridge, NY: Mercury Press, 1997.

Rohen, Johannes. *Functional Morphology: The Dynamic Wholeness of the Human Organism*. Hillsdale, NY: Adonis Press, 2007.

Steiner, Rudolf. *Balance in Teaching*. Great Barrington, MA: SteinerBooks, 2007.

———. *The Child's Changing Consciousness: As the Basis of Pedagogical Practice* (CW 306). Hudson, NY: Anthroposophic Press, 1996.

———. *Cosmic Memory: The Story of Atlantis, Lemuria, and the Division of the Sexes* (CW 11). Great Barrington, MA: SteinerBooks, 1987.

———. *Disease, Karma, and Healing: Spiritual-Scientific Enquiries into the Nature of the Human Being* (CW 107). London: Rudolf Steiner Press, 2013.

———. *Education for Special Needs: The Curative Education Course* (CW 317). London: Rudolf Steiner Press, 2015.

———. *Esoteric Christianity: And the Mission of Christian Rosenkreutz* (CW 130). London: Rudolf Steiner Press, 2001.

———. *Eurythmy Therapy* (CW 315). London: Rudolf Steiner Press, 2009.

———. *Foundations of Esotericism* (CW 93a). London: Rudolf Steiner Press, 1983.

———. *The Foundations of Human Experience*. Hudson, NY: Anthroposophic Press, 1996.

———. *Four Mystery Dramas* (CW 14), rev. ed. Great Barrington, MA: SteinerBooks, 2015.

———. *The Gospel of St. John* (CW 103). New York: Anthroposophic Press, 1962.

———. *Guidance in Esoteric Training: From the Esoteric School* (CW 245). London: Rudolf Steiner Press, 1998.

———. *The Healing Process: Spirit, Nature & Our Bodies* (CW 319). Great Barrington, MA: SteinerBooks, 2010.

———. *Illness and Therapy: Spiritual-Scientific Aspects of Healing* (CW 313). London: Rudolf Steiner Press, 2013.

———. *Introducing Anthroposophical Medicine* (CW 312). Great Barrington, MA: SteinerBooks, 2010.

———. *Intuitive Thinking as a Spiritual Path: A Philosophy of Freedom* (CW 4). Hudson, NY: Anthroposophic Press, 1995.

———. *Man as a Being of Sense and Perception* (CW 206). N. Vancouver: Steiner Book Centre, 1981.

———. *An Occult Physiology* (CW 128). London: Rudolf Steiner Press, 1997.

———. *An Outline of Esoteric Science* (CW 13). Hudson, NY: Anthroposophic Press, 1997.

———. *Physiology and Healing: Treatment, Therapy, and Hygiene* (CW 314). London: Rudolf Steiner Press, 2013.

———. *The Reappearance of Christ in the Etheric: A Collection of Lectures on the Second Coming of Christ*. Great Barrington, MA: SteinerBooks, 2003.

———. *Rosicrucianism and Modern Initiation: Mystery Centres of the Middle Ages* (CW 233a). London: Rudolf Steiner Press, 1982.

———. *Spirit as Sculptor of the Human Organism: The Influence of the Dead* (CW 218). London: Rudolf Steiner Press, 2014.

———. *Start Now! A Book of Soul and Spiritual Exercises*. Great Barrington, MA: SteinerBooks, 2010.

———. *The Temple Legend: Freemasonry and Related Occult Movements: From the Contents of the Esoteric School* (CW 93). London: Rudolf Steiner Press, 2000.

———. *Understanding Healing: Meditative Reflections on Deepening Medicine through Spiritual Science* (CW 316). London: Rudolf Steiner Press, 2013.

———. *The World of the Senses and the World of the Spirit* (CW 134). Rudolf Steiner Press, 2014.

Steiner, Rudolf, and Ita Wegman. *Extending Practical Medicine: Fundamental Principles Based on the Science of the Spirit* (CW 27). London: Rudolf Steiner Press, 1997.

von Laue, Hans-Broder, and Elke E. von Laue. *The Physiology of Eurythmy Therapy*. Edinburgh: Floris Books, 2010.

www.ingramcontent.com/pod-product-compliance
Lightning Source LLC
Chambersburg PA
CBHW030106170426
43198CB00009B/519